Visions for Future Cities

IOP Conference Series: Earth and Environmental Science
Volume 992

Cairo, Egypt
27 – 28 November 2021

ISBN: 978-1-7138-4973-5
ISSN: 1755-1307

Printed from e-media with permission by:

Curran Associates, Inc.
57 Morehouse Lane
Red Hook, NY 12571

Some format issues inherent in the e-media version may also appear in this print version.

Institute of Physics
Dirac House, Temple Back
Bristol BS1 6BE UK

Phone: 44 1 17 929 7481
Fax: 44 1 17 920 0979

techtracking@iop.org

Additional copies of this publication are available from:

Curran Associates, Inc.
57 Morehouse Lane
Red Hook, NY 12571 USA
Phone: 845-758-0400
Fax: 845-758-2633
Email: curran@proceedings.com
Web: www.proceedings.com

TABLE OF CONTENTS

IOP Conf. Series: Earth and Environmental Science **992** (2022) 011001 doi:10.1088/1755-1315/992/1/011001

International Conference
VISIONS FOR FUTURE CITIES
Novel Applications & Technologies
NOVEMBER 2021

Visions for Future Cities
Novel Applications and Technologies
International conference (VFC 2021)

About Conference:

Visions of today's urban planners would be judged by future generations. To create recognition at a glance for future vision, it is of utmost importance that we respect the climate change effects and to obtain sustainability through building, community, and urban levels.

Furthermore, to create a deep truly vision for future urbanism, it is important to be aware of the latest technology themes of digitalization and automation used in the urban planning context. That would ensure the research outcome to be more reliable in the near future to be used and depend on it.

As the world nowadays is changing in accordance with the COVID19 pandemic, the research community should not neglect its effect on the environmental and urban levels. "The new normal" nowadays would surely affect the urban context in the near future.

This conference offers an interdisciplinary gathering in the fields of Environmental urbanism, Sustainable Development, Surveying and Mapping, Urban Design, City Planning, Smart Cities, Renewable Energy, Disaster Management, and Climate Change scenarios on the urban morphology, and many others.

This event is jointly organized by the Cairo Higher Institute (CHI) , ARCH space International academic group and BIM Arabia , in collaboration with the international experts for the enrichment of research community in the field of environmental and urban design and city planning. The event is supported by IOP Publishing and the Journal of Architecture, Arts and Humanistic Science in an attempt to enhance current research and innovation for future and further promote an exchange of knowledge.

CONFERENCE MAIN THEMES:

1. Digitalization and Automation in Architecture and Urbanism.
2. Performance and Assessment in Architecture and Urbanism.
3. New Innovations and Integrated Technologies in Architecture and Urban Design.
4. Urban Ecology and Green Infrastructure.
5. Climatic Changes and Urban Adaptability.
6. Environmental Urbanism, Sustainability and Green technologies

IOP Conf. Series: Earth and Environmental Science **992** (2022) 011001 doi:10.1088/1755-1315/992/1/011001

International Conference
VISIONS FOR FUTURE CITIES
Novel Applications & Technologies
NOVEMBER 2021

7. Futuristic Visions of Cities.
8. Future Buildings, Materials, and Applications.
9. Post-Covid Urban Researches

Date: 27 Nov: 28 Nov 2021
Location: **Safer Hotel, Cairo**

The conference was held in Cairo on 27 and 28 November at Safir Hotel in Dokki. The conference was held with the hybrid systems and the owner of the scientific activities of the conference, two workshops. In addition to the main speakers, they are:

Prof. Kamel Mahadin ,	The American University, Jordan
Prof. Zainab Faisal,	Benha University, Egypt
Prof. Dr. Ahmed Haroun,	Cairo Higher Institute for Engineering, Egypt
Prof. Dr. Alexander Kader,	University of Anhalt, Germany
Eng. Moaz Abu Zeid	Founder Verform, UAE
Mr. Yazan Mahadin	CEO Mahadin Association, Jordan

35 submitted papers for the conference. 26 papers accepted. 20 research papers in English Language, 6 research papers in Arabic Language. Attendee on Conference Avenue was 95 attendees, online was 70 attendees

Tributes to colleagues

The organizing committee thanks all academic partners, colleagues and researchers for their active and positive participation.

Sponsor acknowledgement

The organizing committee thanks Arch Space Academic Group the sponsoring company for their continued support of the conference before, during and after the conference, hoping for continued cooperation between us

IOP Conf. Series: Earth and Environmental Science **992** (2022) 011001 doi:10.1088/1755-1315/992/1/011001

International Conference
VISIONS FOR FUTURE CITIES
Novel Applications & Technologies
NOVEMBER 2021

Conference Board

Conference Chair: Prof. Adel Yassen, Former Dean, Environmental Institute, Ain Shams University , EGY

Head of Scientific Committee: Prof. Ehab Okba, Fayoum University, EGY

Executive Director: Assoc Prof . Ahmed Haron, CHI, EGY

Scientific Committee

Head of Committee:

Prof. Dr. Ehab Okba	**Fayoum University EGY**

Members (Alphabetical):

Prof. Dr. Abd Al Kareem Hassan	Islamic University PL
Prof. Dr. Ahmed Hosny Radwan	BUE EGY
Prof. Dr. Ahmed Mohamed Soliman	Alexandria university EGY
Prof. Dr. Ahmed Youssry	Cairo University EGY
Prof. Dr. Ahmed Rashed	BUE EGY
Prof. Dr. Ahmed Shetawy	6 OCT University EGY
Prof. Dr. Ali Abu Ghanimeh	Jordan university JOR
Prof. Dr. Ali Bahamam	King Saud University KSA
Prof. Dr. Ashraf M. Salama	Strathclyde University UK
Prof. Dr. Ayman Hassan	Cairo University EGY
Prof. Dr. Ayman Othman	BUE EGY
Prof. Dr. Essam Aldin Mohmed	Assiut University EGY
Prof. Dr. Galal Abada	Ain Shams University EGY
Prof. Dr. Gamal Elkholy	Ain Shams University EGY
Prof. Dr. Ghada Farouk	Ain Shams University EGY
Prof. Dr. Heba Allah Khalil	Cairo University EGY
Prof. Dr. Hesham Arif	MSA University EGY
Prof. Dr. Hesham Gabr	Cairo University EGY
Prof. Dr. Hesham Saudi	Alexandria University EGY
Prof. Dr. Hossam Bakr Khalil	BUE EGY
Prof. Dr. Hossam Ibrahim	HBRC EGY
Prof. Dr.Islam Nazmy Soliman	Benha University EGY
Prof. Dr. Jamal Alian	King Saud University KSA
Prof. Dr. Khaled Abd Al Hady	Benha University EGY
Prof. Dr. Khaled Dwidar	BUE EGY
Prof. Dr. Marwa A. Khalifa	Ain Shams University EGY
Prof. Dr. Mostafa Al Shinawy	HBRC EGY
Prof. Dr. Mohamed Al Baramelgy	Cairo University EGY
Prof. Dr. Mohamed Abd Al Aziz	Al Azhar University EGY
Prof. Dr. Mohamed Abo Al Magd	Al Azhar University EGY
Prof. Dr. Mohamed Alaa Mandour	Helwan University EGY
Prof. Dr Mohamed Gamal	University of Nottingham UK

IOP Conf. Series: Earth and Environmental Science **992** (2022) 011001 doi:10.1088/1755-1315/992/1/011001

International Conference
VISIONS FOR FUTURE CITIES
Novel Applications & Technologies
NOVEMBER 2021

Prof. Dr. Mohamed Reda	Cairo University EGY
Prof. Dr Mohamed Shawky	Um Alqurra University KSA
Prof. Dr. Mahmoud Taha	Helwan University EGY
Prof. Dr Mohga Embaby	Fayoum University EGY
Prof. Dr. Shreef Al Attar	Fayoum University EGY
Prof. Dr. Yasser Moharam	Galala University EGY
Prof. Dr. Reham Momtaz	MTI University EGY
Prof. Dr. Wael Yousef	Al Azhar University EGY
Prof. Dr. Walaa Nour	Tanta University EGY
Prof. Dr. Zeinab Faisal	Benha University EGY
Assoc. Prof. Ahmed Haron	CHI EGY
Assoc. Prof. Alexander Kader	Anhalt Honchschule GRM
Assoc. Prof. Amira Elnokaly	University of Nottingham UK
Assoc. Prof. Mai Wahba Madkour	Tanta university EGY
Assoc. Prof. Walaa S.E. Ismaeel	BUE EGY

Organizing Committee
Head of Committee:

Assoc. Prof. Ahmed Haron	**CHI EGY**

Members:

Dr. Hayam Omayer	CHI EGY
Dr. Eman Hatem	CHI EGY
Dr. Aya Alkholy	CHI EGY
L.A Shimaa Magdy	CHI EGY
L.A Dalia Abdelfatah	CHI EGY
L.A Manar Mohamed	CHI EGY
L.A Haidy Ehab	CHI EGY
T.A Panssee tarek	CHI EGY
T.A Rola Mohamed	CHI EGY
T.A Darin Tarek	CHI EGY
T.A Mohamed Ahmed	CHI EGY
T.A Fatma Bahgat	CHI EGY
T.A Aya Mohamed ELgendy	CHI EGY
T.A Heba Gaber	CHI EGY

IOP Conf. Series: Earth and Environmental Science **992** (2022) 011002 doi:10.1088/1755-1315/992/1/011002

Peer Review Statement

All papers published in this volume have been reviewed through processes administered by the Editors. Reviews were conducted by expert referees to the professional and scientific standards expected of a proceedings journal published by IOP Publishing.

- **Type of peer review:** Double Anonymous
- **Conference submission management system:** Morressier
- **Number of submissions received:** 18
- **Number of submissions sent for review:** 17
- **Number of submissions accepted:** 12
- **Acceptance Rate (Submissions Accepted / Submissions Received × 100):** 66.7
- **Average number of reviews per paper:** 2.8
- **Total number of reviewers involved:** 47
- **Contact person for queries:**
  Name: Ahmed
  Email: info.vfc2021@gmail.com
  Affiliation: Arch Space International Group - Academic

IOP Conf. Series: Earth and Environmental Science **992** (2022) 012001 doi:10.1088/1755-1315/992/1/012001

Impact of the Geometric Form of the Building Envelopes on The efficiency of Natural Lighting in the Office Space.

Mohammad A Y Bhai [1], M Abdelkader [2], A Neseem [3], A Mustafa [4]

[1] Postgraduate Student, Department of Architecture, Ain Shams University, Cairo, Egypt. E- mail: mabahi_78@yahoo.com

[2] Professor of Architecture and Environmental Control, Ain Shams University, Egypt.

[3] Associate Professor, Department of Architecture, Ain Shams University, Egypt.

[4] Lecturer, Department of Architecture, Ain Shams University, Egypt.

Abstract: Envelopes play an important role in Office Buildings.They affect visual comfort for the user Driven by architectural trends and the need to maximize natural light In order to realize required visual comfort conditions. advanced glazing systems, Geometric formation of the outer building envelope proper self- shading properties and control and lighting controls should be studied comprehensively Optimized fenestration system design may improve exploitation of daylight. and it can be controlled to harvest maximum daylight thus reaching the maximum depth naturally illuminated. The paper explains the design relationship between both configuration of office space and between the system of the glazing and the transmission of natural light inside the space, in terms of the design characteristics of the office space (proportions, dimensions), in order to achieve visual comfort to improve the efficiency of the environmental performance of Office Buildings. This paper discusses the study of the environmental effects of the glazing façades of office buildings and their impact on the efficiency of natural lighting and visual comfort. The paper relies on simulation using, the Grasshopper program, in analyzing of the building's Envelopes and its external elements and the architectural space, in order to find the proper form for improving visual performance.

Keywords: Visual Comfort, Glazing Systems, Natural Light, Glazing Envelopes, Simulation Program, Office Buildings.

1. Introduction

The architectural model is important in terms of depth, proportions, shape and type of space (shallow space and deep space), which mainly contribute to the thermal transmission and natural lighting in the interior space, affecting the quality of environmental efficiency (visual comfort). The architectural model needs to be determined in terms of the depth and proportions of the internal spaces of multi-level office buildings [1].

The glass envelopes that separate indoor environment from the outside environments act as a barrier against extreme temperatures, solar radiation and light transmission. However the problem is using large glass surfaces for windows irrespective of the climatic considerations in hot dry regions [2].

The paper focuses on the building envelopes and its effect on the characteristics of the architectural space (depth) in office buildings, in order to reach the optimum level of natural lighting and visual comfort in the internal environment.

2. Study hypotheses

a) The natural lighting can be improved within the office buildings through the characteristics of the geometric form and the glazing of the buildings envelope.
b) Determining the design characteristics of the façade of the office space at an early stage in the designing, according to the quality, properties and of the glass area in relation to the depth and dimensions of the office space.
c) Possibility of benefit from the geometric configuration of the properties of the building's glass envelope in improving the efficiency of the visual environment of internal environment.

3. Problem Definition

a) Lack of knowledge of the natural lighting criteria to improve the visual environment within the architectural space.
b) Lack of knowledge of the properties of the geometric configuration of the building's facade, in order to achieve the best possible environmental performance and energy saving.
c) The envelopes of modern office buildings are implemented with glazing characteristics with less light transmittance than the permissible.
d) No methodology to determines the relationship between the properties and systems of glazing and the architectural properties of the office space in terms of the depth and proportions of the space, to improve the efficiency of natural lighting.

4. Objectives of the study

4.1. The main objective

a) Improve the efficiency of the visual environment in interior office spaces.

4.2. Procedural objectives

a) Highlighting the importance of daylight in creating a comfortable visual environment suitable for the practice of office activity.
b) Simulation analysis of the efficiency and quality of natural daylight levels within the office architectural space.
c) An Analysis of the performance of daylight by two main criteria:
   - Annual Sun Exposure (ASE)
   - Spatial Daylight Autonomy (SDA)

5. Study methodology

5.1. First: Theoretical Study

This part deals with the study of the interrelationships between the characteristics of the office space and of its glazed outer shell, and their impact on achieving visual comfort and natural lighting levels through the glass element used in the envelope of the office buildings [3].
It investigates the impact of these properties on achieving visual comfort and getting to know the simulation programs used to measure natural lighting standards [4].

5.2. Second: Applied Study

Simulating the performance of natural daylight for a Case study in Egypt:
Methodologies used include literature review to collect data concerning different aspects of glass Properties and glazing system of Glass facades, comparative analysis study of visual environment simulation outcome to assign the optimum design for Glass building envelopes [2].
The following methodology was adopted through the proposed design for the hypothetical office space, and the steps are: [5].

a) *Design criteria for the office space and its outer envelope...*
   - Systems and characteristics of glass windows and the glazing element for the building envelope.
   - Characteristics of the architectural form of the outer building envelope.
   - The depth of the office space (shallow or medium) depth.
   - The form of the office space and its relation to the office work environment.

B) *Environmental criteria...* [6].
- Visual comfort: by assessing the visual design criteria and considerations which must be taken into account when designing the natural daylight in the office building.
- The integration of interior artificial and natural daylight within the office space.

5.3. Field of applied study

The study includes the use of the programs (Rhinoceros, Grasshopper, Energy Plus, and Plugins) in identifying the elements and components of the proposed approach to improve the visual performance of natural daylight of the office space at the design stage:

a) Study the geometric form of the building envelope (tilting of the glass surface).
b) Study the properties of the glazing system of the building's envelope to achieve Visual Performance efficiency in the architectural space.
c) Study the Impact of building orientation on comfort levels for natural lighting of office space in buildings.
d) Determine the maximum naturally lit depth of the office space.
e) Determine the lowest depth that can be achieved by integrating artificial lighting and natural Lighting.

6. Properties of the geometric form of the building envelope and its effect on visual comfort

Architectural form has a great the impact on the development of the building's envelope design in terms of transmission of natural lighting and environmental control [7].

6.1. Visual comfort:

It is a condition where human eyes receive a suitable amount of light without visual strain, which strongly relate to the illumination levels inside the office space [6].

6.1.1. Main parameters of visual comfort

The visual environment must allow seeing objects clearly without strain in pleasantly toned surroundings. Where the parameters of visual comfort are: the level of illumination of visual tasks, the harmonious distribution of light within a space, the ratios of illuminance within a building, exterior and internal shadows elements, good color interpretation, pleasant tones of light, and the absence of glare [6].

6.1.2 Glare

Glare is Visual discomfort that interferes with visual performance. Two main types of glare exist, direct and indirect (reflected), and each can have very detrimental effects on Visual comfort [8].

Direct glare is caused by a Direct light source in the field of the view that is sufficiently bright to cause annoyance, discomfort, or loss in visual performance. Reflection of light sources on the glossy surface usually cause indirect glare [6].

6.1.3 Visual comfort Vs human performance

Lighting quality is one of the determinants of human performance in the indoor environment. Where Various some studies have been conducted to compare between the effects of different lighting conditions and levels on health, productivity, and well-being and alertness level, office Work performance [6].

IOP Conf. Series: Earth and Environmental Science **992** (2022) 012001 doi:10.1088/1755-1315/992/1/012001

6.2. Glass envelope

Glass envelope is an important element in buildings and the main function of it is to protect the internal from the external environment conditions, so energy-efficient building Researchers have started to develop their researches based in this field .

Today, energy usage in buildings have become an important factor hence building construction materials such as glass and modern glass technology should be considered accordingly early Design Stages [9].

a) *Quality of Light ...*
   Uniformity of light: Light uniformity represents how light spreads over a task area. Uniform light distribution helps in avoiding the risk of visual discomfort caused by various eye adaptations to different levels of light, whether it is very bright or low light which leads to visual stress [10].

b) *The Amount of light...*
   An adequate amount of natural daylight provides good that allows users to accomplish their work tasks. Higher or lower light levels can cause visual discomfort.
   Illuminance is the physical quantity usually used to quantify the light amount that reaches a specific point of a work surface [10].

c) *Light quality...*
   Light uniformity: This means how light is spread over an important area. Even light distribution helps avoid the risk of visual disturbance from highly lit or low-lit areas, which leads to Visual discomfort [10].

7. Description of the design proposal

The study case is a hypothetical office space model of the city of Egypt. Simulation modeling will be conducted during the four seasons of the year and during daylight hours. The case study is located in Cairo (longitude 31.5^{O}, latitude 29.8^{O} northeast). It has been assumed that all spaces have an open horizon and there are no external obstacles.

8. Daylighting Optimization

Study was concerned with daylighting optimization encompassing both external and internal screen configurations simultaneously. The analysis was conducted using Daylight Performance Metrics (DDPM); the IES metrics (IESNA, 2012) which give benchmark levels for the pass or fail metrics for Natural lighting. The first metric (sDA300/50%) gives an indication of Natural lighting adequacy inside the space, where a minimum illuminance of 300 lux is meant to be reached at 50% of the occupied hours across at least 55% of the space. The second metric (ASE 1000/ 250 hr) indicates excessive sunlight exposure when receiving direct sunlight of 1000 lux for more than 250 hours. This should not exceed 10% of the space area; optimal cases should reach at least 75% sDA and a maximum value of 10% ASE to avoid possible visual discomfort. Illuminance levels were calculated on the work plane set on a level 0.8 m from the ground and was divided into a grid of 0.25 m*0.25 m. The time WORK was set from 9 am to 5 pm. Radiance parameters used for sDA and ASE calculations were adjusted according to the (LEED V4 & IES) [11].

Table 1. Natural Daylight standards in accordance with LEED V4 & IES requirements [12].

Requirements (LEED V4 & (IES)		Required conditions	Nature lighting standards
Acceptable Acceptable Favorite	10% or less 7% or less 3% or less of the area	1,000 lux for at least 250 annual business hours	Annual Sun Exposure (ASE)
Acceptable Favorite	55% or more 75% or more of the area	300 lux for at least 50% of annual working hours	Spatial Daylight Autonomy (SDA)

Table 2. Architectural specifications for the office space model- (Source: Researcher)

Architectural characteristics of the space	Space elements
North - East - South - West	Orientation
Office	Use
4 * 6 * 2.8 m (24 m^2 space area)	Space dimensions
Net height = 2.8m Total height of the turn = 3.2m False ceiling = 40cm	Space height
Characteristics of the glass window	
Net window height = 2.5m Window width = 4m = width of the space Building/concrete depth = 0.30m	Window dimensions
90% calculating the upper inclined part (solar cells) 80% glass with light transmission	WWR
6mm double glazing 20 mm gap Glass color (gray) Light transmission 40%,50%	Glass properties
 Located in Cairo (longitude 31.5 ^{O}, latitude 29.8 ^{O} northeast).	Geolocation coordinates
0.80 cm 0.25*0.25 cm	Level of work Divided level of work
During the seasons of the year During daily work hours from 9am to 5pm	Time/period

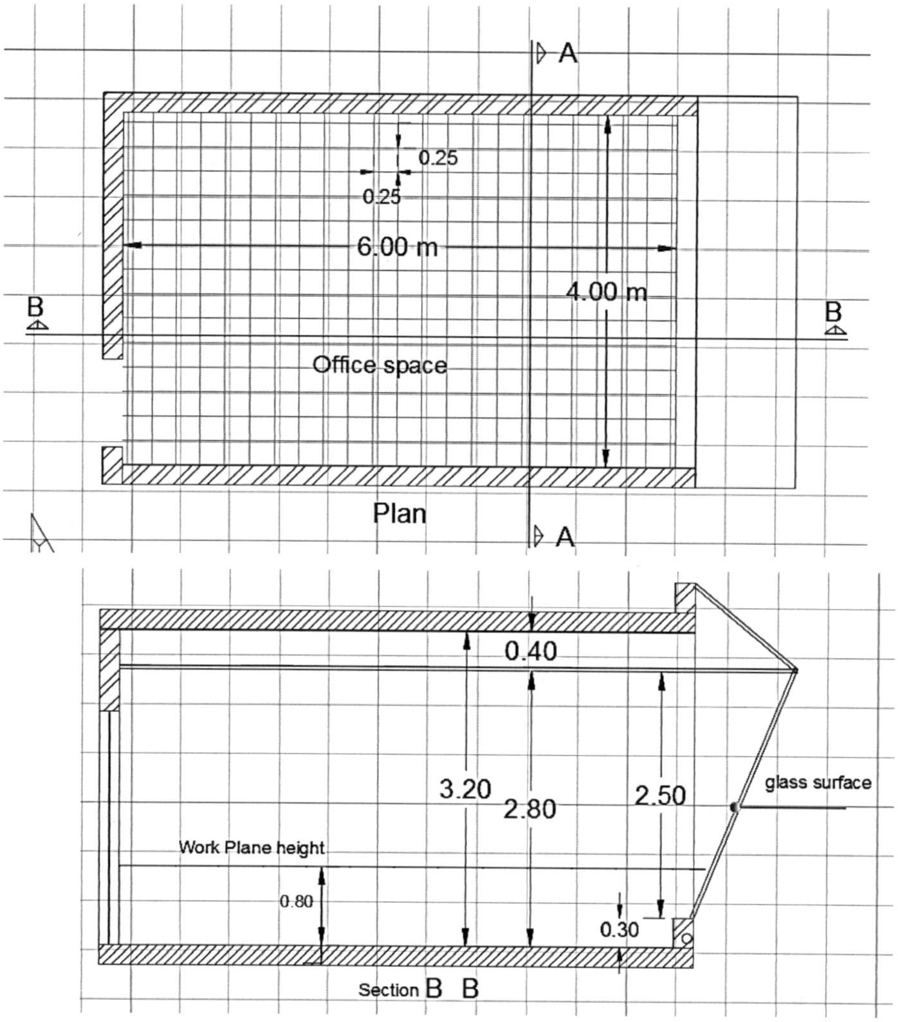

Figure 1. Architectural characteristics of the office space- DESIGN PROPOSAL (Source: Researcher)

9. The applied study

9.1. The applied study simulation

In this study, the performance of natural lighting will be evaluated through simulation program (Rhino - Grasshopper), for the previously mentioned hypothetical model in two successive analytical stages.

9.2. The (design proposal)

Simulate the study model with the characteristics of the geometrical formation of the building façade and to define specific items for the characteristics of the glass and to determine the best percentage of glass transmission for natural lighting of the glass, according to the previously defined natural lighting criteria.

(DESIGN PROPOSAL)
Case study
A study of the characteristics of the geometric form of the building envelope and simulation of glass transmittance for natural lighting of glass

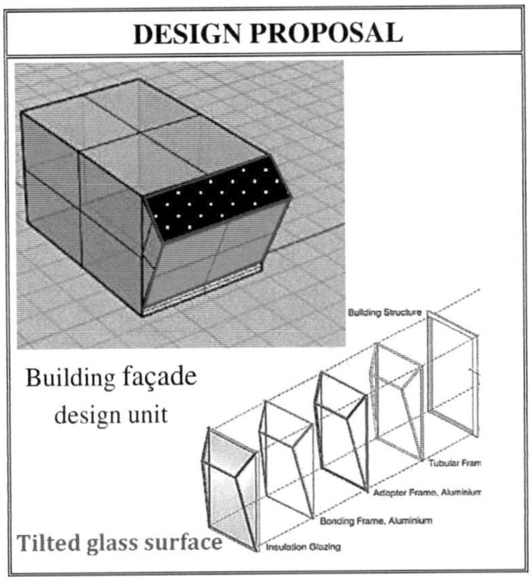

Figure 2. Case study and Design Proposal- (Source: Researcher)

The second stage includes simulating the study model to add characteristics of the glass and determining the optimum natural light transmission of the glass, according to the criteria for natural lighting (SDA and ASE criteria), then accessing the results of the optimization process and its impact on lighting efficiency and energy consumption.

10. Study proposal idea

The idea of research is based on the avoiding the traditional pattern of the vertical glazing system commonly used in buildings facades, the proposal for improving performance includes the concept of tilting the building façade to a specific angle (self-shading), to maximize the amount of natural lighting, as an architectural prerequisite.

a) If the outer shell is tilted at an angle of 10°, this is a small angle of inclination that does not affect the process of improving visual.

b) If the outer envelope is tilted at an angle of 30°, which is a very large inclination, the building façade protrudes 1.42m, contrary to Law 119 of 2009 regulating construction work, which states that the maximum permissible projection is 1.25 m.

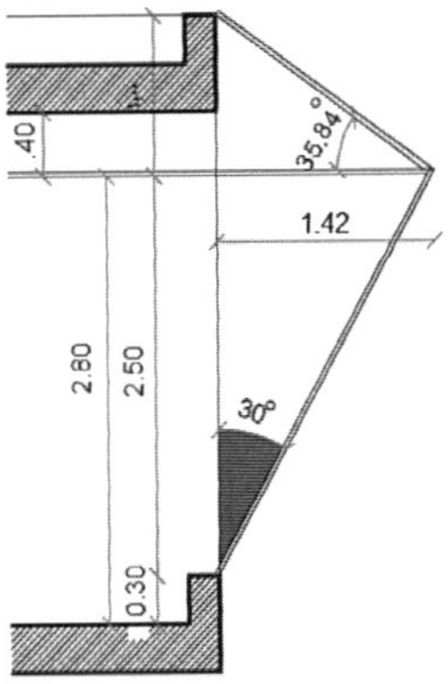

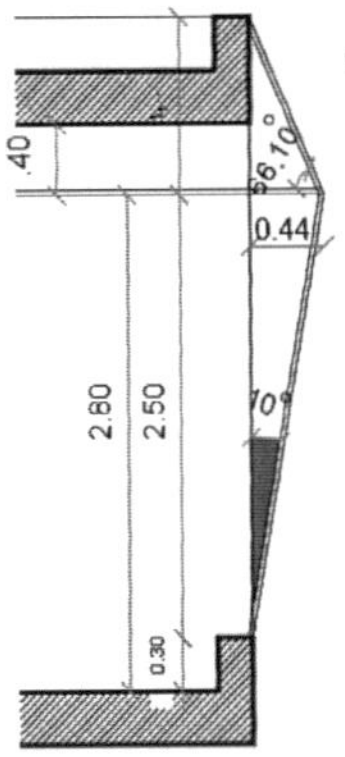

Figure 3. Tilting the outer shell by **30°** **Figure 4.** Tilting the outer shell by **10°**
(Source: Researcher)

c) Where work is done at this stage to tilt the outer façade by 20°, the building façade protrudes about 1 m, as it does not exceed 1.25 m, according to Law 119 of 2009 regulating construction work.

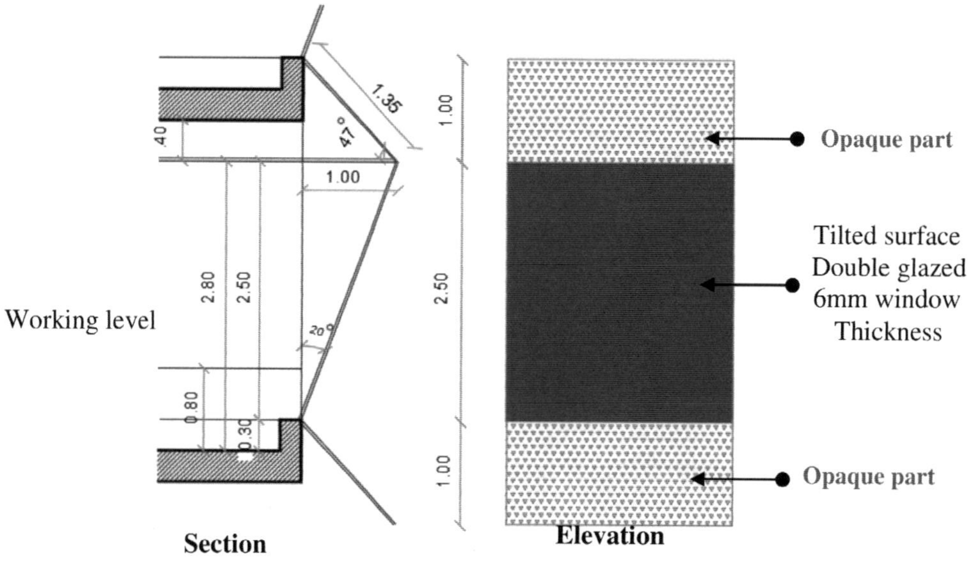

Figure 5. Tilting the outer shell by **20** (Source: Researcher)

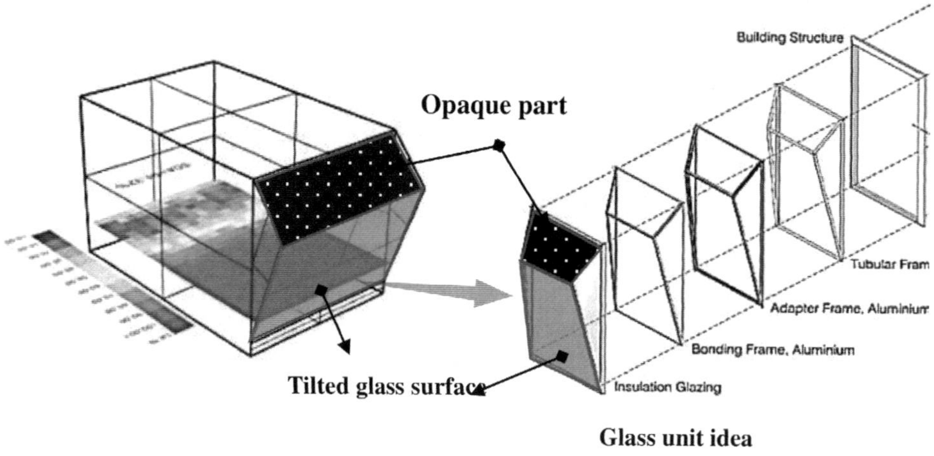

Figure 6. Simulation program inputs for the study model – Grasshopper-Rhino, (Source: Researcher)

11. Analysis of simulation

a) Southern façade

Table 3. Analysis of the results of SDA and ASE standards for 40 %transmission and interface tilt angle of 20° - (Source: Researcher)

The light transmittance is LT = 40% and the interface is tilted at an angle = 20°		Geographic orientation
Annual Sun Exposure ASE	Spatial Daylight Autonomy-SDA	
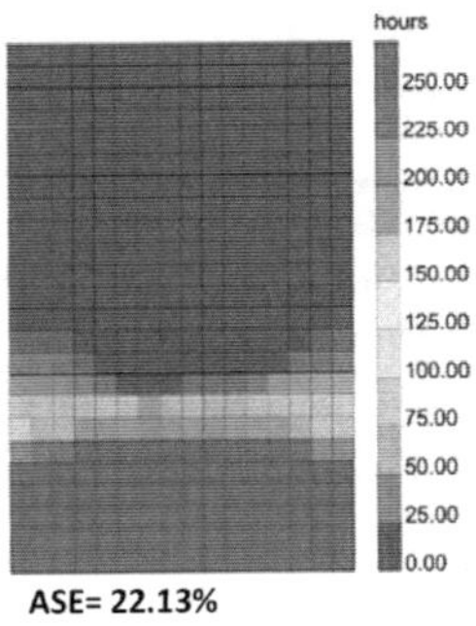	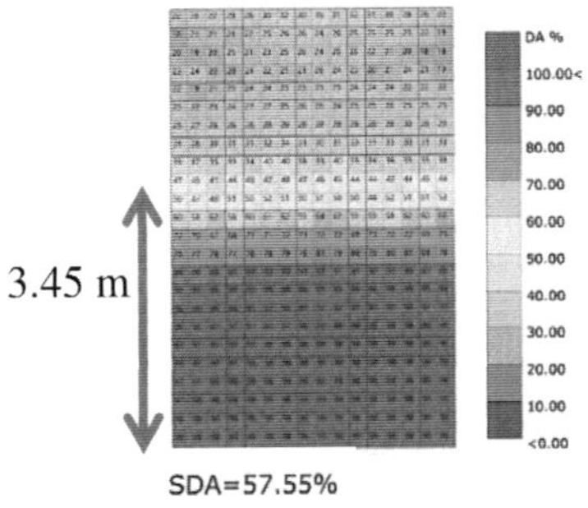	Southern facade

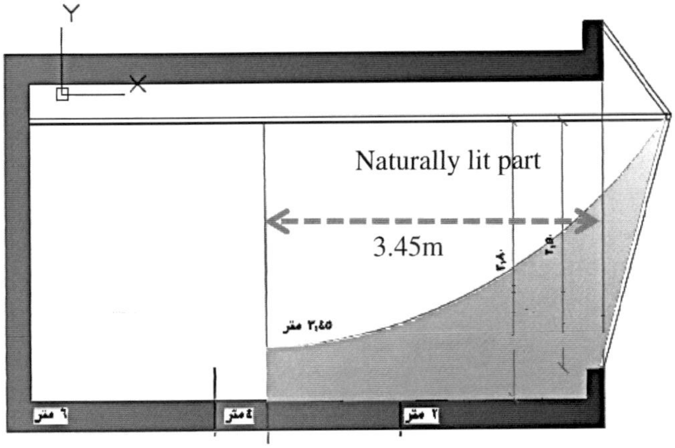

Figure 7. Section

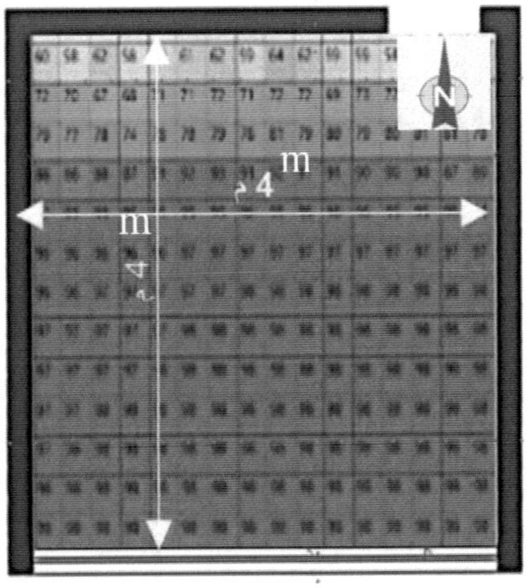

Figure 8. Plan

(Source: Researcher)

The Glass of light transmission (40%) achieved a value of SDA is (57.55%). This is an acceptable percentage according to LEED V4 requirements. Meanwhile, the glass with a light transmission rate of (40%) achieved a significant improvement in the ASE standard in accordance with the requirements of LEED V4, and the Illuminating Engineering Society (IES), which recorded a value of (22.13%) of the room's surface.

A light transmission rate of 40% was achieved, with a depth of no more than 4m (Maximum architectural depth) for the office space, naturally lit (Depth Lighted naturally = 4*3.45 m).

IOP Conf. Series: Earth and Environmental Science **992** (2022) 012001 doi:10.1088/1755-1315/992/1/012001

b) Eastern façade

Table 4. Analysis of the results of the SDA and ASE criteria for 50% transmission and an interface tilt angle = 20 ^{O} - (Source: Researcher)

The light transmittance is LT = **50%** and the interface is tilted at an angle = 20°		Geographic orientation
Annual Sun Exposure ASE	Spatial Daylight Autonomy-SDA	
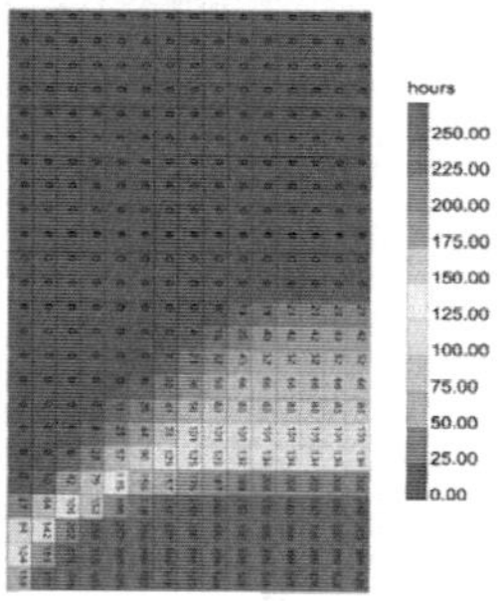 ASE =13.05%	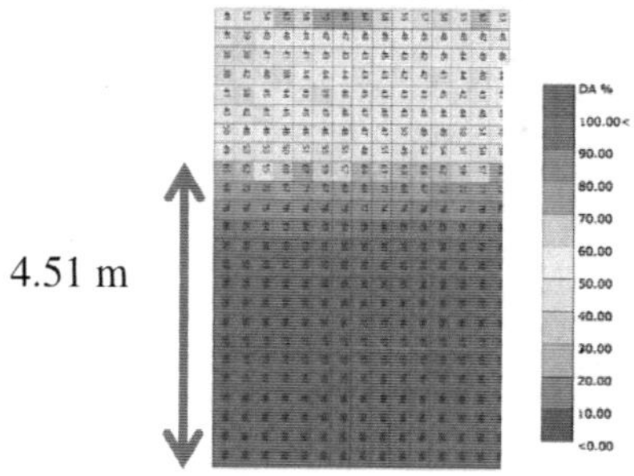 SDA=75.28%	Eastern facade

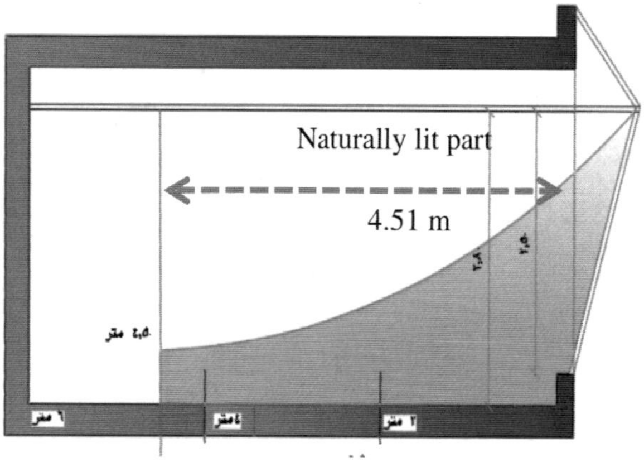

Figure 9. Section

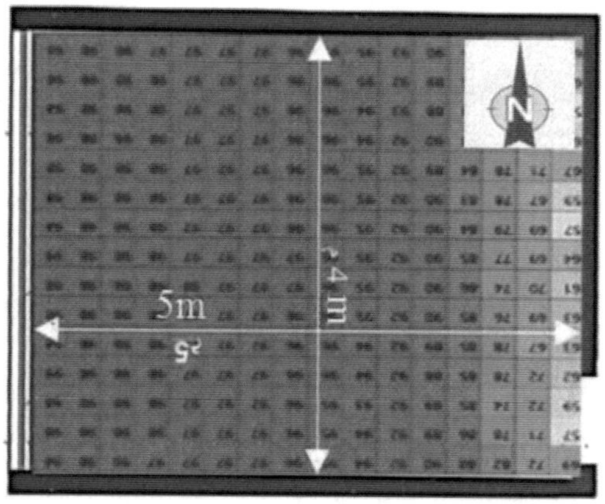

Figure 10 .Plan

(Source: Researcher)

The glass of light transmission of (50%), achieved a value of SDA is (75.28%). This is a good percentage according to LEED V4 requirements. Meanwhile, the glass with a light transmission rate of (50%) achieved a significant improvement in ASE, with a value of (13.05%) of the room's surface, which is a good percentage according to the requirements of LEED V4 and the IES .

A light transmission rate of 50% was achieved, with a depth of no more than = 5m (Maximum architectural depth) for the office space, naturally lit, (Depth Lighted naturally = 4* 4.5 m).

c) Western façade

Table 5. Analysis of the results of the SDA and ASE criteria for 50% transmission and an interface tilt angle = 20 ^{O} - (Source: Researcher)

The light transmittance is LT = 50% and the interface is tilted at an angle = 20°		Geographic orientation
Annual Sun Exposure ASE	Spatial Daylight Autonomy- SDA	
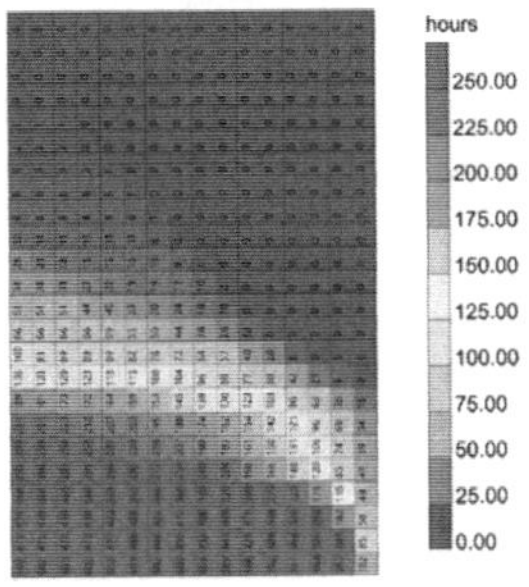 ASE =17.44%	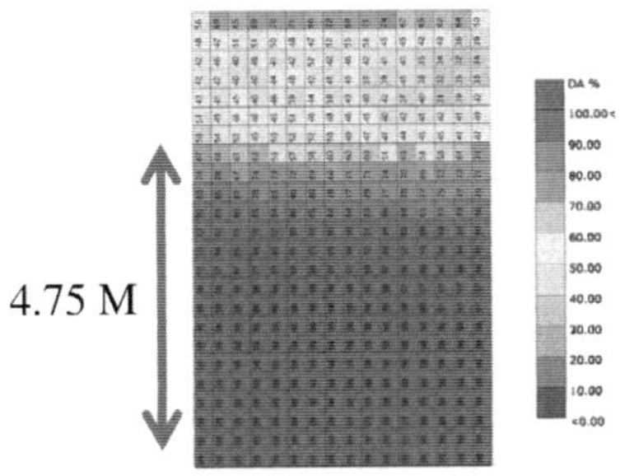 SDA=79.17%	Western façade

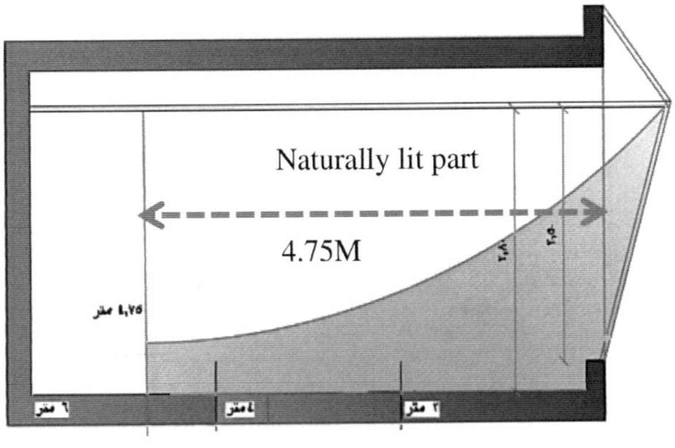

Figure11. Section

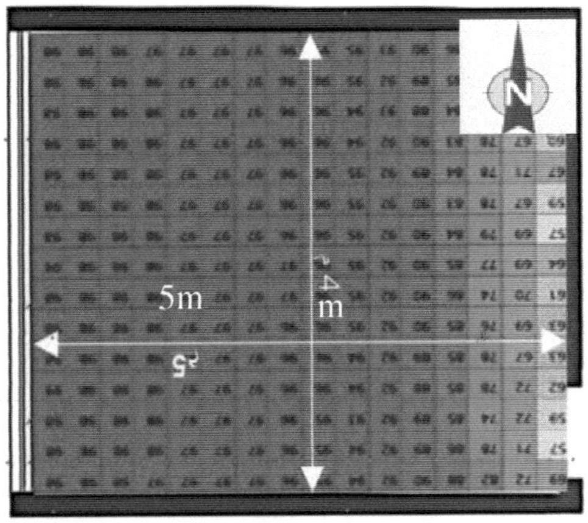

Figure 12. Plan

(Source: Researcher)

The glass of light transmission of (50%), achieved a value of SDA is (79.17%). This is a very good percentage, according to the requirements of LEED V4. Meanwhile, the glass with a light transmission rate of (50%) achieved a significant improvement in the ASE standard, which recorded a value of (17.44%) on the room's surface, which is a good percentage according to the requirements of LEED V4 and the IES.

A light transmission of 50% was achieved, with a depth of no more than = 5m (Maximum architectural depth) for the office space, naturally lit, (Depth Lighted naturally = 4 * 4.75 m).

D- Northern façade

Table 6. Analysis of the results of the SDA and ASE criteria for 50% transmission and an interface tilt angle = 20 ^{O} - (Source: Researcher)

<table>
<tr><th colspan="2">The light transmittance is LT = 50% and the interface is tilted at an angle = 20°</th><th rowspan="2">Geographic orientation</th></tr>
<tr><th>Annual Sun Exposure
ASE</th><th>Spatial Daylight Autonomy-
SDA</th></tr>
<tr><td>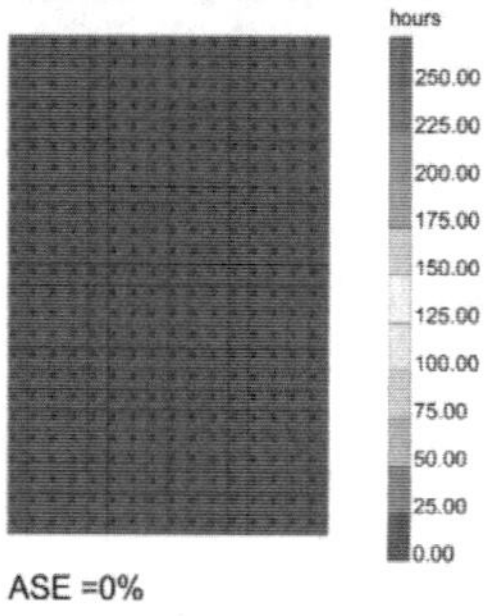
</td><td>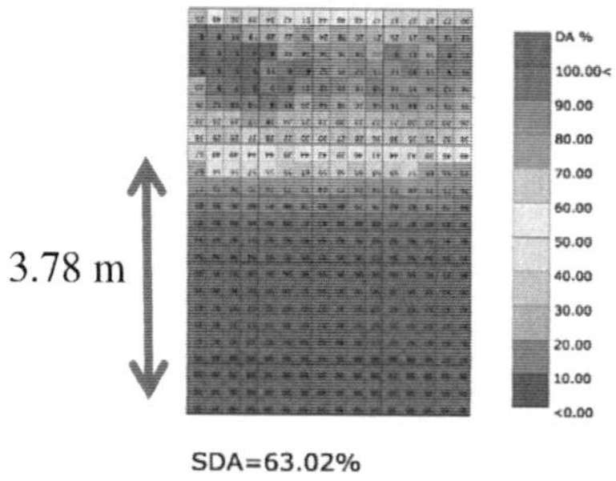
</td><td>Northern façade</td></tr>
</table>

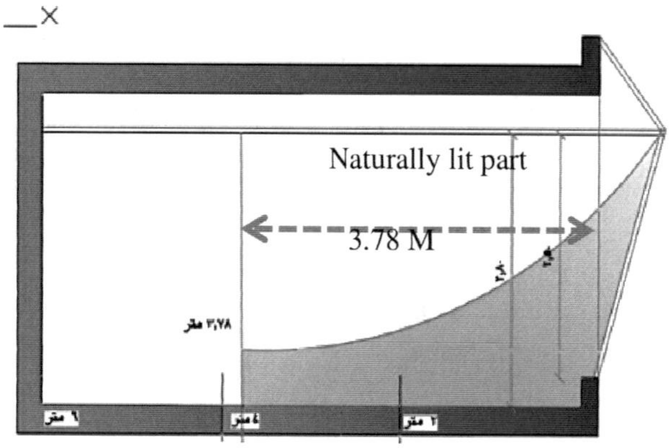

Figure 13. Section

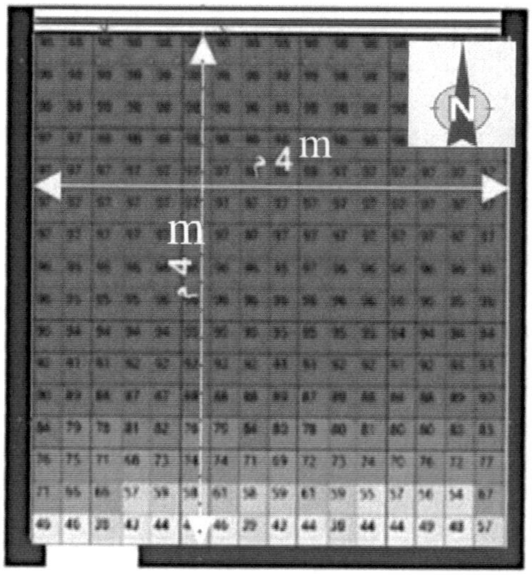

Figure 14. .Plan

(Source: Researcher)

The glass of light transmission (50%) achieved a value of SDA is (63.02%). This is a good percentage according to LEED V4 requirements and the IES. Meanwhile, the glass with a light transmission rate of (50%) met the ASE standard, which recorded 0% for the space area, as the northern façade is not exposed to the sun's direct rays, which is a good percentage according to the requirements of LEED V4 and the IES.

It achieved aA light transmission rate of 50% was achieved, with a depth of no more than = 4m (Maximum architectural depth) for the office space, naturally lit Depth Lighted naturally = 4 * 3.78 m).

12. Analysis of artificial lighting for office space:

Integration of artificial lighting design with natural lighting

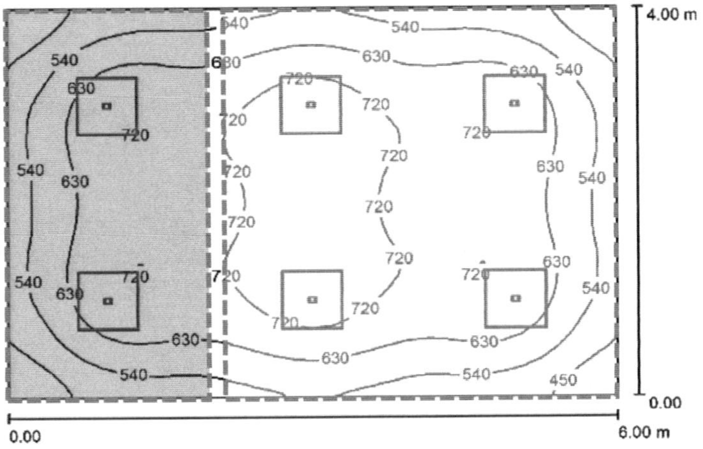

Figure 15. Artificial lighting distribution scheme using LED headlight units (2 units 60 * 60 cm) the power of the flashlight consumption is 35 watts using Dialux program to light for reaching the maximum depth (6 meters) of the architectural office space. (Source: Researcher)

Height of Room: 2.800 m, Mounting Height: 2.500 m, Maintenance factor: 0.80

Surface	ρ [%]	E_{av} [lx]	E_{min} [lx]	E_{max} [lx]	u0
Workplane	/	634	384	785	0.607
Floor	20	515	342	628	0.663
Ceiling	70	124	90	145	0.725
Walls (4)	50	331	95	557	/

Workplane:		UGR	Lengthways-	Across	to luminaire axis
Height:	0.800 m	Left Wall	21	21	
Grid:	32 x 32 Points	Lower Wall	20	20	
Boundary Zone:	0.000 m	(CIE, SHR = 1.00.)			

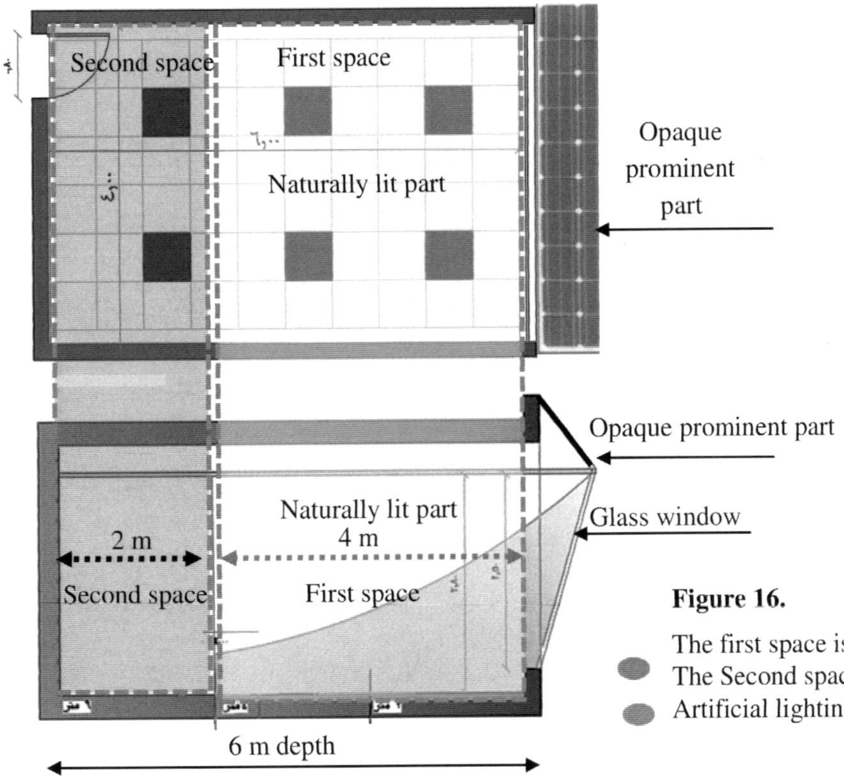

Figure 16.

The first space is naturally lit
The Second space is artificially lit
Artificial lighting units

13. Conclusion and recommendations

Future research is needed to evaluate the overall energy effects of the natural daylight designs. As for modern office buildings, the tilt angle for natural daylight is 20° for the exterior envelope. The architectural design should reflect the geometric form of the envelopes, taking into account the glass and the glazing system used, to achieve maximum visual and environmental comfort from the natural lighting, which helps the designer to achieve the best standards and levels of daylighting to achieve visual comfort.

13.1 Characteristics of the external architectural geometric form

tilting the exterior glass of the building (20°), an appropriate angle in terms of maximizing the transmission of natural light, while improving both the criteria of Spatial Daylight Autonomy (SDA) and Annual Sun Exposure (ASE), with an external protuberance of no more than 1.25m, in line with Law 119 of 2009 for regulating building work.

It is very important to**:**

a) Tilting the outer glass surface at an angle of 20°
b) Determine the window area relative to the outside wall area (WWR)
 (a WWR of 80%), to achieve maximum light transmission
c) Transmittance Transmission coefficient of glass window (VLT)
 (VLT) is 40% as a minimum for south orientation
 (VLT) is 50% as a minimum for (east, west and north) orientation.

13.2. Architectural design characteristics

a) Show the maximum depth of the naturally lit architectural space, according to the analysis of simulating natural lighting criteria Spatial Daylight Autonomy (SDA).
b) Create architectural spaces of appropriate dimensions suitable for carrying out office activity according to the distribution of natural lighting within the space .

13.3. Criteria and levels of natural lighting

a) The SDA should conform to the LEED V4 and IES criteria to achieve visual comfort within the office space.
b) The level of ASE should be reduced to improve visibility and reduce glare from direct sun, in compliance with LEED V4 and IES within the office space.

13.4. Criteria of artificial lighting

a) Achieving integration between natural lighting and artificial lighting for energy saving.
b) Relying on the enhancement of artificial lighting units to achieve the required levels of lighting within the office space reaching the maximum depth (6 meters) of the architectural office space.
c) Increasing the depth of space to achieve large areas of office space to suit all activities and uses.

- It is important to study the architectural formation and sustainable formation, before launching on the architectural design, in addition to continuing with any design work that has already commenced.
- Replacing the Geometric form properties of traditional glazing systems, based on the vertical building façade, with Geometric form properties as one of the design solutions to improve the interior environment natural lighting and energy efficiency.
- Using these criteria SDA, ASE it is possible to achieve the depths of naturally lit designer can rely on during the first design, based on the effectiveness and results of specialized simulation programs, in order to improve natural lighting and internal environmental performance of office space.

IOP Conf. Series: Earth and Environmental Science **992** (2022) 012001 doi:10.1088/1755-1315/992/1/012001

- Achieving the maximum depth of the naturally lit architectural space according to the simulation analysis of natural lighting standards.

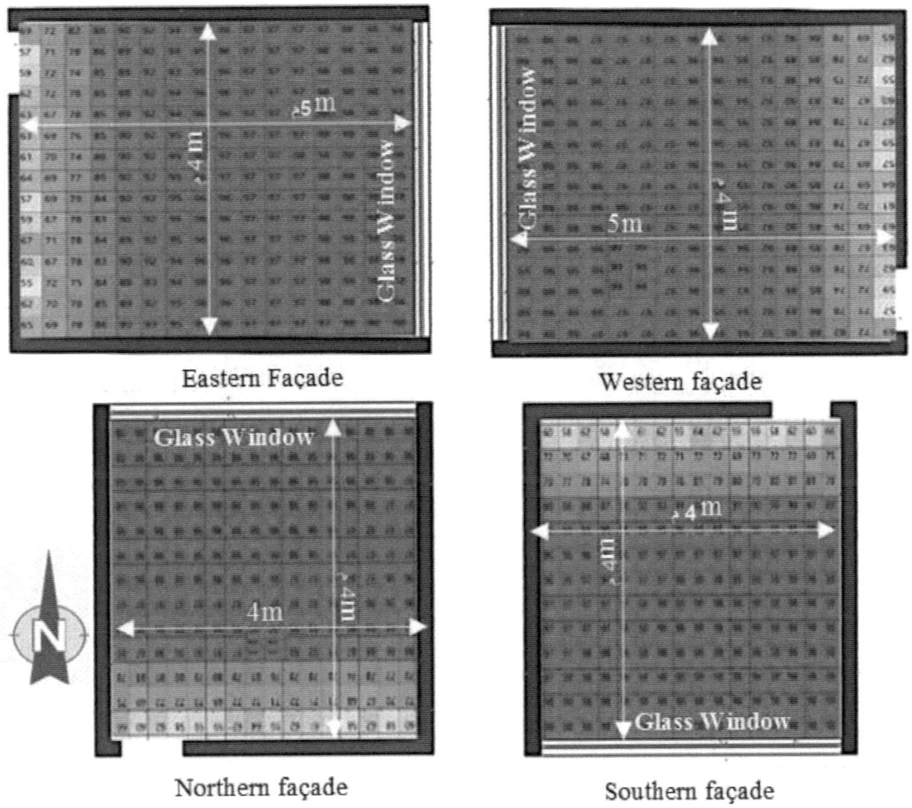

Figure 17. Maximum depth of naturally lit architectural space- (Source: Researcher)

- Southern orientation, (Maximum architectural depth) of 4 meters
- Eastern orientation, (Maximum architectural depth) of 5 meters
- Western orientation, (Maximum architectural depth) of 5 meters
- North orientation, (Maximum architectural depth) of 4 meters

IOP Conf. Series: Earth and Environmental Science **992** (2022) 012001 doi:10.1088/1755-1315/992/1/012001

Table 7. The effectiveness of the side openings in the event that the building envelope is tilted at an angle of inclination of 20° – (Source: Researcher)

Maximum architectural depth of natural lighting	VLT	Orientation	WWR	The angle of inclination of the building envelope
4*4m Realized to natural lighting standards	not less than 40%	Southern	**%80**	An inclination angle of 10° is not effective for improving environmental performance and natural light efficiency
4*٥m Realized to natural lighting standards	not less than 50%	Eastern	**%80**	An inclination angle of **20°** is effective for improving environmental performance and natural light efficiency
4*5 m Realized to natural lighting standards	not less than 50%	Western	**%80**	An inclination angle of 30° is effective for improving environmental performance and natural light efficiency This is in violation of Law 119 of 2009 regulating construction work to exceed the prominence of 1.42 m
4*4m Realized to natural lighting standards	not less than 50%	North		

References:

[1] Santa Raymond & Roger Cunliffe (1997), *Tomorrow's Office – Creating Effective & Humane Interiors , E & FN Spon* , (London), P 129.

[2] Mohamed Ahmed & Mohamed Anwar Fikry (2019), *Impact of glass facades on internal environment of buildings in hot arid zone*, (Alexandria University), (Alexandria Engineering Journal), P. 1

[3] Gada abo zed (2016), *interior design of open plan office and its impact on individual performance* , PhD, College of Fine Arts, (Minia University).

[4] Reinhart, C. F., & Wienold, J. (2011). *The daylighting dashboard–A Simulation-based design analysis for daylight spaces*, Building and Environment, 46(2), Page 386-396.

[5] Yusuf, H. O. (2012), *The impact of digital-computational design on the architectural design process* , (Cairo University). *Doctoral dissertation*, Master's Thesis, (University of Salford).

[6] M. David et al, (2011), *Assessment of the thermal and visual efficiency of solar shades, build. Environ*. 46 (7) 1489–1496, https//: doi.org/10.1016/j.buildenv.2011.01.022

[7] Dalia Samir Mikhael (2005), *the Impact of Technological Development on Architectural Formation*, Master Thesis, Faculty of Engineering, (Cairo University).

[8] Tobias Schulze, Ursula Eicker, (2013) *Controlled natural ventilation for energy efficient buildings Energy Build*, https://doi.org/10.1016/j.enbuild.2012.07.044

[9] Sustainable Bioclimatic Second-Skin Fac¸ades Save Eergy, https://www.imoa.info/molybdenum-uses/molybdenum-gradestainless-steels/architecture/bioclimatic-facades.php..

[10] Carlucci, S., Causone, F., De Rosa, F., & Pagliano, L. (2015), *a review of indices for assessing visual comfort with a view to their use in optimization processes to support building integrated design*. P. 1018.

[11] Hanan Sabry, Ahmed Atef Faggal, (2017), *External Versus Internal Solar Screen: Simulation Analysis for Optimal Daylighting and Energy Savings in an Office Space*, Conference Paper, (Ain shams University, Cairo).

[12] Measuring Daylight: Dynamic Daylighting Metrics & What They Mean for Designers, URL: http://sefaira.com/resources/measuring-daylight-dynamicdaylighting-etrics-what-they-mean-for-designers.

IOP Conf. Series: Earth and Environmental Science **992** (2022) 012002 doi:10.1088/1755-1315/992/1/012002

The Evolution of Life Cycle Assessment Approach: *A Review of Past and Future Prospects*

Aliaa AMahmoud [1,2], Mohamed M El Attar [3], and Alaa MEleishy [4]

[1]Department of Architecture, Faculty of Engineering, Mansoura University, Mansoura, Egypt

[2]Teaching assistant, Department of Architecture, Mansoura College Academy, Mansoura 35511, Egypt, Email: aliaaabbas1989@gmail.com

[3] Professor, Department of Architecture, British University in Egypt. New Cairo 11835, Egypt.

[4] Assoc. Professor, Department of Architecture, Mansoura University, Mansoura 35511, Egypt.

Abstract. There is consensus evidence that the thousands of years known earth's capability to sustain life has been seriously eroded, especially since the industrial revolution. Modern forms of human existence related to prompt economic development have given a massive share in the permanent degradation of the plant, its ecosystem, resources, and quality of life of its inhabitants. Consecutively, there is increasing environmental awareness in material use and its environmental impacts, resulting in the proliferation of environmental assessment tools such as Life Cycle Assessment approaches. Ever since the first approach was proposed in the 1960s, LCA has undergone robust development in both approaches and applications. It is now a well-known and widely used tool across the industry, academia, and policy. So, this paper aims to introduce the LCA hotspots in history with a highlight on the main five periods: Early years period, conception period, standardization period, elaboration period, concept period. It is vital to address the growth in the futuristic aspiration of the new methodology related to integrating Building Information Modelling (BIM) with the LCA procedure to provide real-time results in the early design stages.

1. Introduction

Sustainable development was developed its official definition in the Brundtland report 1987 to eliminate the negative consequences of rising only the quality of life for the present generation. It embraces the idea of having the ability to meet the current needs without threatening the future capability to satisfy their own needs [1]. Based on the fact that we are rapidly depleting non-renewable resources; energy, and raw materials, the consciousness of sustainability has rapidly increased worldwide in the last decade. The awareness is with the target of avoiding the environmental impacts occurred by different types of human activities. Activities with harmful ecological effects vary from one industry to another. However, it is stated that the building industry is the most contributor to Green House Gases emissions (GHG). [2]

The ACE sector (Architecture, Construction and Engineering) has intensive attention as the building industry robustly influences the environment. It is considered to have the largest share in the natural resources drain and the foremost reason for undesirable impacts such as air and water pollution, solid waste, deforestation, toxic wastes, health hazards, global warming, and other harmful effects [3]. It is considered as the most leading cause of carbon emissions and energy consumption as it represents 36% of worldwide energy use and nearly 40% of total direct and indirect CO2 gas emissions [4]. Global warming currently is one of the highest profiles which is a consequence of the long-term accumulation of GHG emissions; CO2, CH4, N2O, etc., in the higher layer of the atmosphere. [5]

GHG emissions impacts have the most consideration from researchers and policymakers, although it is just one of a wide range of impacts [6]. So, it becomes clear that there is a need for a holistic perspective of materials to evaluate all the environmental impacts throughout its whole life cycle. Thus, to evaluate a material's total impacts, all life cycle stages as shown in (Fig1) must be considered, from material extraction, manufacturing, construction, use, including operations, maintenance, demolition, and disposal. [7]. The main reason for considering all types of environmental issues is to avoid what is called "Burden Shifting". It happens when the improvement in a specific stage of the material life cycle concerning one kind of environmental impact can lead unintentionally to a rise in the other types of impacts for the same stage or different life cycle stages. [8].

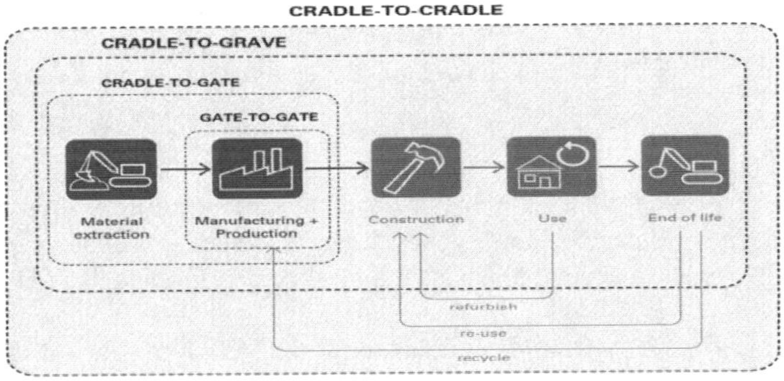

Fig 1: Life Cycle stages of a building with different approaches [7]

In that context, Life Cycle Assessment (LCA) has been indicated to be an applicable emerging approach to evaluate products and materials concerning their environmental performance. LCA has undergone a robust development in approaches and applications since it was conceived in the 1960s [9]. LCA promises to strengthen the architectural decision-making by investigating the environmental impact of buildings materials and concluding to techniques to decrease these impacts [10]. Even though the LCA method provides a quantifiable measurement of environmental impact, awareness of the strength and weaknesses points of LCA is essentially required to ensure the ability to understand the LCA procedure. [7]

From the early beginning to now, methodological development of (LCA) has still undergone with attention to the new approaches and tools to support the seamless procedure. So, this paper aims to give a comprehensive historical overview of the (LCA) development throughout the years, focusing on the scope of approaches development, application, international standardization, and broadcasting. In addition, the paper also addresses the futuristic aspiration of the new methodology related to integrating Building Information Modelling (BIM) with the (LCA) procedure.

2. Life Cycle Assessment (LCA): Philosophical perspective

The LC allegory is driven from the biology field. In biology, the "LC" term refers to the organism constant sequence of changes from one initial form, as a gamete, to the development of the same form again for the cycle to be repetitive" [11].

2.1. Philosophy of construction- economy development

The economy has adopted a linear philosophy for quite a long time. The Linear Economy concept in the construction sector as shown in (Fig 2) is based on raw material take-make-use-dispose. This concept seeks a short-term consumption of materials resources and if it remains in the linear model, the reserves available for materials will be depleted in a few decades. It has resulted in environmental issues such as climate change, natural capital loss, and health problems, etc.. [12]

IOP Conf. Series: Earth and Environmental Science **992** (2022) 012002 doi:10.1088/1755-1315/992/1/012002

Fig 2: The construction – economy concept (take-make-use-dispose)[12]

Despite the linear economy success in generating material fortune in the industrial countries up to the 20th century, it has shown a weakness in the new millennium [13]. Thus, as shown in (Fig 3) Circular Economy philosophy (CE) is swept over in the construction sector which emerges as a counterpoint to the Linear one. In CE, products and raw materials are designed to be reusable to maintain and enhance natural resources and minimize risk by managing limited resources and renewable loops. [14]

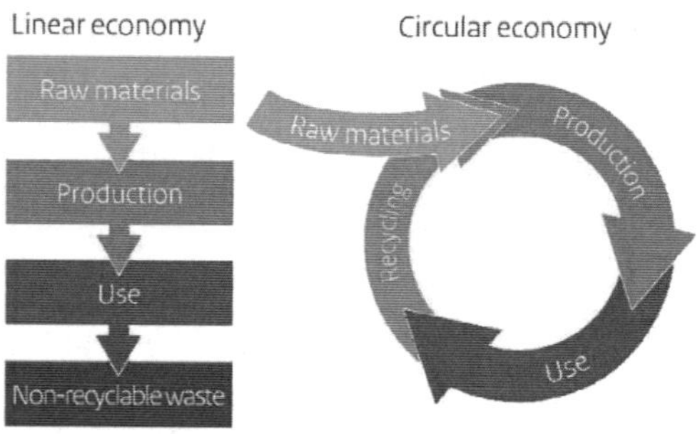

Fig 3: From Linear to Circular Economy concept. [14]

According to the European Academies Science Advisory Council (EASAC), there are critiques of the CE related to the recycling process. Achieving the closed-loop of a material that already caused pollution, waste through its life cycle requires energy and resources. This consumption is used for the recovery and recycling process which increases in a nonlinear manner as the percentage of recycled material rises. [15]

Furthermore, in addition to mitigating environmental impact, companies using this concept can reduce besides the environmental impacts reduce the production cost, the dependence on natural raw materials, and also reduce the environmental. To understand the concept of CE, a philosophy of Life Cycle Thinking (LCT) has been engaged in the construction field to have a comprehensive perception of the material life cycle. [14]

2.2. *The Philosophy of Life Cycle Thinking (LCT)*

In a simplified way, every material has its own lifetime process called the Life Cycle. It has been proven that a material's life cycle results in a wide range of environmental impacts. Therefore, the Life Cycle Thinking (LCT) concept refers to taking into consideration the upstream and downstream of the material in a holistic perspective. [16]

Furthermore, the LCT concept deals with all material life stages as shown in (Fig 4)from the raw materials and the energy used to manufacture the material to the packaging, distribution, use, maintenance, and finally recycling, reuse, or final disposal. Each one consumes energy, water, and resources and results in waste, emissions, and a specific environmental impact. This concept has its target to mitigate this consumption and promote material performance. [17]

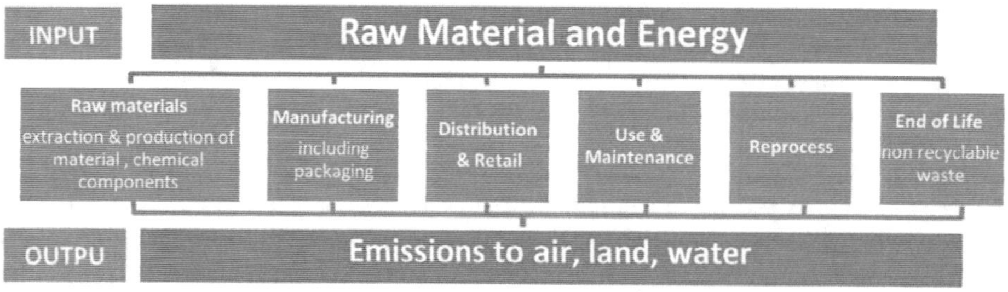

Fig 4: Life Cycle Thinking (LCT) approach [17]

Furthermore, LCT philosophy is equivalent to the "6 RE philosophy" which includes as shown in (Fig 5), a) rethink about the material or product and its functions, b) reduce the depletion of the energy and natural recourse, c) replace the material with harmful effect with the environmental option, d) recycle the material to make a closed-loop and reduce the waste, e) reuse the existed material, f) repair the material. [16]

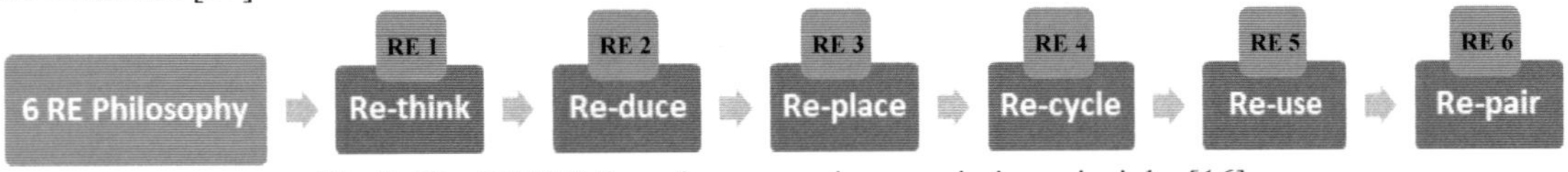

Fig 5: The 6 RE Philosophy concerning eco-design principle. [16]

LCT is a mindful approach that goes beyond the traditional linear concept by analyzing a product or material life cycle (from the cradle to the grave) and engaging the three sustainability aspects in the approach as shown in (Fig 6). These aspects are the environmental aspect embodied in the LCA as a tool, the social aspect represented in Social Life Cycle Assessment (SLCA), and the economic aspect embodied in Life Cycle Costing (LCC). Keeping all three in balance is the key to make sustainable products or materials. [18]

Consequently, the ultimate target of LCT is to enhance the environmental and Socio-Economic performance throughout its life cycle. In detail, it aims to help companies and customers to be more aware of how their decisions impact the environment from a holistic perspective and make better alternative decisions to mitigate environmental impacts. [18]

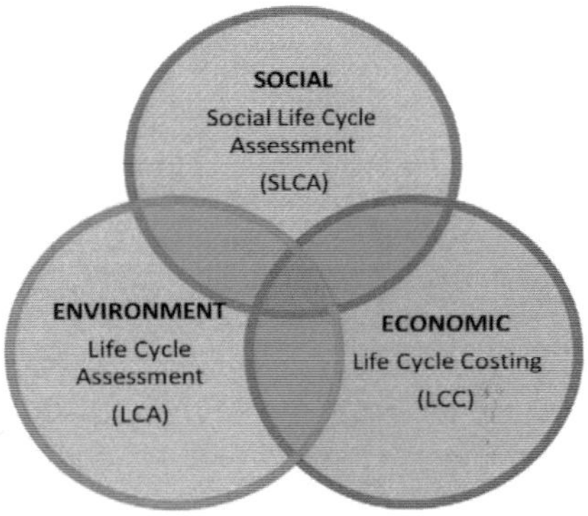

Fig 6: Life cycle thinking key to every aspect of sustainability. [18]

2.3. LCA formal definition

Throughout the years, LCA has had different attempts to be officially defined as a new approach. The first initial definition was introduced in 1990 by the SETAC workgroup as a comprehensive tool for assessing the environmental burdens related to a product, process, or activity by analyzing the energy, materials used, and wastes generated. The assessment involves the product's whole life cycle from extraction of raw materials, manufacturing, transportation and distribution, use/reuse/maintenance, recycling to final disposal [19]. Furthermore, ISO 14040 also defined LCA as a systematic approach for evaluating the inputs and outputs of materials and energy related to a process or product to analyze the potential environmental impacts caused throughout its life cycle. [9]

3. Life Cycle Assessment (LCA): Historical perspective

The development of Life-Cycle approaches has been driven by concerns related to the environmental pollution and the depletion of materials and energy. Ever since the first methodology was proposed in the 1960s, LCA has undergone robust evolution in both methodology and applications. So, it is now a well-known and widely used tool across the industry, academia, and policy [8].

From its start in the 1960s to the broad-spread in just over 60 years,LCA has passed through different stages of development and adoption as shown in (Fig 7); a) the early years period (1960-1970), b) Conception period (1970-1990), c) Standardization period (1990-2000), d) Elaborations' period (2000-2010), e) Concept broadening period (2010-till now) [20]. The major transition events in the LCA history are summarized chronologically in (Fig 8) and will be discussed in detail in the below sections. [20]

IOP Conf. Series: Earth and Environmental Science **992** (2022) 012002 doi:10.1088/1755-1315/992/1/012002

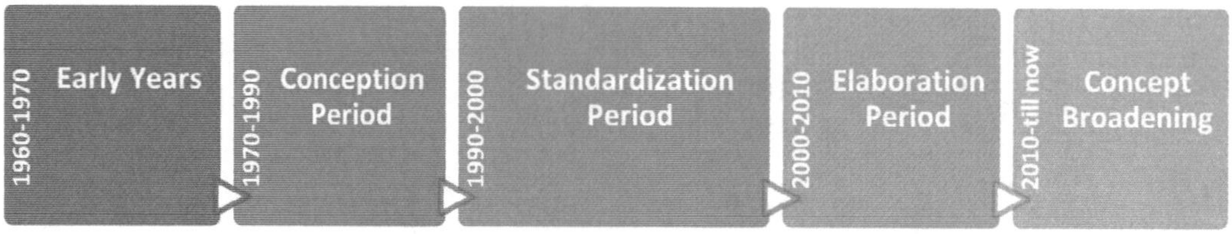

Fig 7: Life Cycle Assessment (LCA) development over 60 years

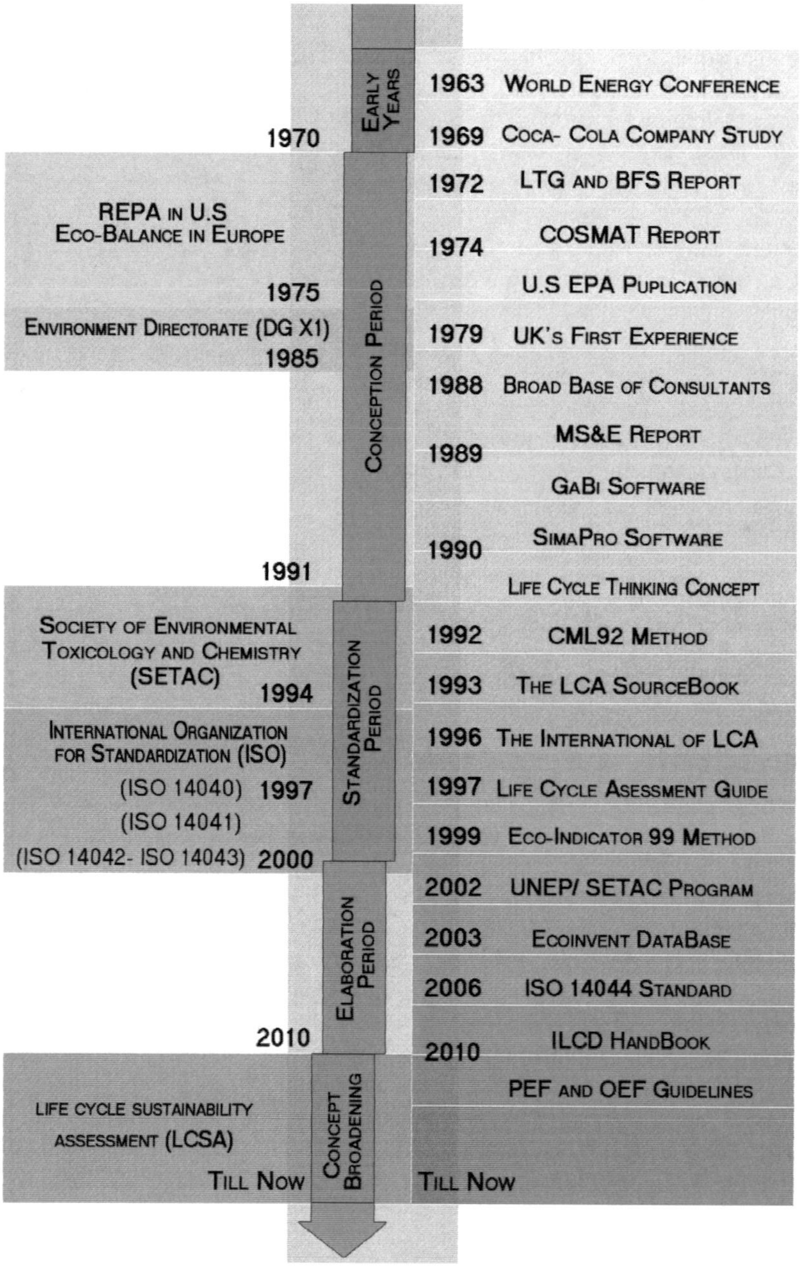

Fig 8: The timeline of the transition events in LCA history

IOP Conf. Series: Earth and Environmental Science **992** (2022) 012002 doi:10.1088/1755-1315/992/1/012002

3.1. The early years period (1960-1970)

LCA had its initial steps in the 1960s as shown in (Fig 9) in finding ways to take into account the energy and materials' resources throughout the product manufacturing. Various of these early LCA attempts weren't published as they were either commercially sensitive or internal company reports never planned for public distribution. The Coca-Cola Company was the first manufacturing firm to address LCA related to its beverage containers. As in 1969, the company funded research to evaluate the consumption of material resources and the environmental impacts. [21]

1963: World Energy Conference

- **At World Energy Conference,** Harold Smith declare his compution of accumulative energy demands for the manufacturing of chemical intermediates and products which represent one of the first publication of its type.

1969: The Coca-Cola Company Study

- During of the energy crisis, researchers started an internal study for **The Coca-Cola Company** establishing the groundwork for the current approach of life cycle inventory analysis in the U.S. The raw materials, fuels, and environmental loadings from the manufacturing processes for each individual beverage container were calculated in order to determine that had the least enviornmetal effect and consumed the fewest natural resources.

Fig 9: The transition events in the early period (1960-1970) [21]

3.2. The conception period (1970-1990)

The period 1970-1990 embodied the LCA conception spread with broadly diverging approaches, terminologies, and results. During the 1970s and 1980s, LCAs were carried out utilizing various methodologies without a common theoretical framework, as illustrated in (Fig 10). Firms used LCA to satisfy the substantiate market demands. In 1974, U.S Environmental Protection Agency (US. EPA) performed the first peer-reviewed study to inform regulation on the packaging. The results varied considerably for the same product, prohibiting LCA from becoming a more widely recognized and used analytical method. [22]

1972: The LTG and A BFS Report

- Several worldwide studies published in **The Limits to Growth (LTG)** and **A Blueprint for Survival (BFS)** resulted in inaccurate projections of the effects of growing populations on finite raw materials and energy resources demand. During this time, a dozen studies were conducted to determine the economic and environmental consequences of various alternatives.

1974: COSMAT Report

- **Committee on the Survey of Materials Science and Engineering (COSMAT)** was announced at a critical juncture in the modern industrial history. This happend in 1973 just after the OPEC oil crisis, the freeing of currencies from gold standard, the birth of modern econmic triad. In the spirit of the OCMAT study, LCA approaches and techniques are being widely developed.

1974: US EPA Report

- **The U.S Environmental Protection Agency (US EPA)** performed the first peer-reviewed study in 1974 to inform regulation on the packaging. Even for the same product, the results varied greatly, preventing LCA from becoming a more widely recognized and used analytical method.

1979: UK's first experience

- A handbook of industial energy analysis published by Boustead and Hancock was The UK's first experience of the life-cycle perspective.

Fig 10: The transition events in the 1970s period. [22]

Alongside these specific events, there are two transition events in the LCA development through the 1970s. The process of predicting environmental burdens and material resources used is known as a Resource and Environmental Profile Analysis (REPA) in the U.S and Eco-balance in Europe. Between 1970 and 1975, approximately 15 REPAs were performed. Within this time, protocol or methodology for conducting these studies was developed, including several assumptions. During these years, EPA and major industry representatives had considerably reviewed these assumptions and techniques. [8]

From 1975 to the early 1980s, the European Commission established the Environment Directorate (DG X1) to develop methodologies similar to those used in the United States, which piqued European interest. In addition to standardizing pollution regulations in Europe, in 1985, (DG X1) published the Liquid Food Container Directive to examine the energy and raw materials consumption and the solid waste creation of liquid food containers. [8]

During the 1980s, a wide range of consultants and researchers as shown in (Fig11) has been developed the methodology. The need for dedicated LCA software has been identified as a result of the demand to go beyond the inventory to impact assessment. Around 1990, the first version of SimaPro and GaBi, two extensively used programs, were announced. Until the 1990s, LCT wasn't a well-known concept outside of the packing industry. [23]

1988: Broad Base of Consultants and Researchers

- The approach has been refined and expanded by a wide range of resarchers and consultants from all around the world. LCA methodolgy has gone to a new level of development due to the necessity to overtake the inventory to impact assessment.

1989: MS&E Report

- The "materials tetrahedron" concept was the work base in the **The Materials Science and Engineering (MS&E)** report that focused mainly on the using of materials.

1989: GaBi Software

- **GaBi Software**, the most comnon progrom was releaed in 1989. It was one of the primaly commercial software tools to take part into the market-leading LCA tool and data base.

1990: SimaPro Software

- **SimaPro**, another widely used commercial LCA software, was released in its first version

1990: Life Cycle Thinking Concept

- The International Chamber of Commerce nad the Society of Environmental Toxicology and Chemistry (SETAC) organized conferences and working groups to improve and standaedized LCA methodlolgy. **Life Cycle Thinkijg (LCT)** was Until the 1990s, LCT wasn't a well-known concept outside of the packing industry.

Fig11: The transition events in the 1980s period. [23]

3.3. The standardization period (1990-2000)

The number of seminars, guides, and handbooks in the 1990s reflected the outstanding global coordination activities growth. From 1991 to 1994, a cooperation amongst LCA specialists was established through the North American and European branches of the Society of Environmental Toxicology and Chemistry (SETAC). This collaboration was resulted continual development and standardization of LCA framework and methodology. One of the most important outcomes of this coloration was the "Code of Practice". [5]

Moreover, the International Organization for Standardization (ISO) has been engaged in LCA since 1994. Parallel to the focus of the SETAC working group on improving and harmonizing methodologies, ISO took the formal mission of methods and procedures standardization. Despite the lack of standardized methods, the standard met industry requirements for using LCA for product development. Over the next seven years, four standards were released based on the standard development: a) the principles and framework (ISO 14040) in 1997, b) the goal and scope definition (ISO 14041) in 1998, c) the life cycle impact assessment (ISO 14042) in 2000 and d) the life cycle interpretation (ISO 14043) in 2000. [5]

Besides these two pivotal events, there were several attempts to standardize the LCA methodology as shown in (Fig 12). For instance, the CML92 method created by the University of Leiden in the Netherlands, and the Eco-indicator method, the first endpoint assessment technique and still the furthermost widely used in LCA. [23]

1992: CML92 Method

- **CML92** was the first impact assessment approach to encompass a broad range of midpoint impact categories. It was released by the Institute of Environmental Sciences at Leiden University in the Netherlands in 1992.

1993: The LCA Sourcebook

- **The LCA Sourcebook: A European Guide to Life Cycle Assessment** was published in 1993 and is the most comprehensive international overview of LCA activities to date.

1996: The International Journal of Life Cycle Assessment

- **The International Journal of Life Cycle Assessment**, the first academic journal dedicated entirely to LCA, was founded in 1996. The number of papers published in this and other journals has risen dramatically, from less than 100 in 1998 to more than 1300 in 2013.

1997: Life Cycle Assessment Guide

- **The book "Life-cycle assessment (LCA) – a guide to approaches, experiences and information sources"**, was published by European Environment Agency (EEA) with the aim of supporting business and readers to navigate the LCA maze and discover the tight tools for their applications.

1999: Eco-indicator 99 Method

- The approach that was introduced by the Dutch **Eco-indicator 99** methodology released in 1999 with a more scientific approach.

Fig 12: The transition events in the 1990s period. [23]

IOP Conf. Series: Earth and Environmental Science **992** (2022) 012002 doi:10.1088/1755-1315/992/1/012002

3.4. The elaboration period (2000-2010)

From the beginning of the 21st century, the focus on LCA has been rapidly growing. As a result of its qualitative approach to framework, impact assessment, and data quality, it has become a vital decision support tool and broadly used methodology. So, the period 2000-2010, as shown in (Fig 13) can be defined as the decade of elaboration. This period has a broad experience in establishing recognized databases encompassing all industrial sectors and aiming for reliable standards, such as the well-known Ecoinvent Database in 2003. [22]

In 2002, the Life Cycle Initiative, international cooperation, was established by the United Nations Environment Program (UNEP) and (SETAC). It has three programs: a) the Life Cycle Management (LCM) program to raise consciousness and enhances decision-making skills by providing information; b) the Life Cycle Inventory (LCI) program with the aim of improving global access to transparent, high-quality life cycle data; and c) the Life Cycle Impact Assessment (LCIA) program to improves the quality and global reach of life cycle indicators. [22]

2002: UNEP/SETAC

- the Life Cycle Initiative, an international cooperation, was established by **the United Nations Environment Program (UNEP) and the Society of Environmental Toxicology and Chemistry (SETAC)**. The Initiative has three programs: a) the Life Cycle Management **(LCM)** program; b) the Life Cycle Inventory **(LCI)** program; and c) the Life Cycle Impact Assessment **(LCIA)** program.

2003: Ecoinvent Database

- The resource usage and emissions of a same industrial process might varied significantly amongst databases due to variances in data standards and quality. In 2003, the first **Ecoinvent database (version 1.01)** was released, spanning all industrial sectors and striving for standardized data standards and quality.

2006: ISO 14044 Standard

- **The ISO 14044 standard** detailed the criteria and guidelines in a 2006 modification, without modifying any of the standards' requirements. The ISO 14040 series of Environmental Management standards deals with LCA methodology, but there are also standards and technical guidance reports on LCA applications such as eco-design (ISO 14062, ISO 14006), environmental performance communication (ISO 14020 series on ecolabels and ISO 14063), and greenhouse gas reporting and reduction in (ISO 14064).

2010: ILCD Handbook

- **The International Life Cycle Data System (ILCD)** was developed by The European Commission includes a database of life cycle inventory data as well as a set of methodology guidelines. The ILCD project also included a comparison of all known LCIA methodologies, comparing their approaches to evaluating the various midpoint and endpoint impact categories and determining a recommended practice for each impact category.

Fig 13: The transition events in the 2000s period. [22]

3.5. The concept broadening period (2010-till now)

Lately, in the twenty-first century, A life cycle sustainability assessment (LCSA) framework has emerged with a target of considering the environmental, social, and economic dimensions of sustainability. LCA concept was originally intended to solely address the ecological component. In contrast, as a result of these recent developments, LCA has evolved from a purely environmental LCA to a further holistic LCSA evaluation. The following is a formal LCSA framework: [24]

LCSA = LCA + LCC + S-LCA;

The International Society for Industrial Ecology (ISIE) established a section in 2011 to concentrate on the LCA as it was and on LCSA as a developed framework. Until the end of 2018, there are 258 papers available including 146 case study papers. The number of publications has increased as shown in (Fig 14) between 2016 and 2018 compared to before 2016. Although, a wide range of obstacles related to LCSA remains ambiguous, e.g., defining sustainability aspects and the need for multi-criteria decision analysis. [25]

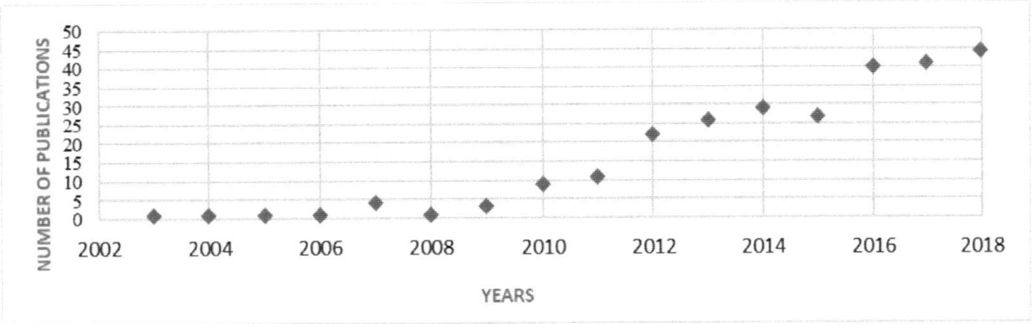

Fig 14: Overview of LCSA publications until end of 2018. [25]

Since 2012, there have been extensively ambitious opportunities to make use of the emerging digitization in the ACE sector to overcome the environmental challenges with Building Information Modelling (BIM) adoption. It enables the comprehensive exchange of building information among various experts. [26]. So, integrating the BIM method into the LCA calculation can provide instant indicators of the effects of the design decisions on the building's environmental efficiency. From the architect's perspective, this can boost the decision-making in the early design stage and improve the efficiency, time-consuming, and complexity and hence participate in broader dissemination of the LCA implementation. [27]

Substantial evolution was observed concerning evolving BIM-LCA studies to improve the LC of buildings throughout the different design stages. As shown in (Fig 15), compared with the initial studies in 2012 and the studies and applications reached in 2015, the BIM-LCA has progressively turned from performing LCA in the early design stage to assessing the LCA based on a comprehensive design that reducing manual activities and time. [28]

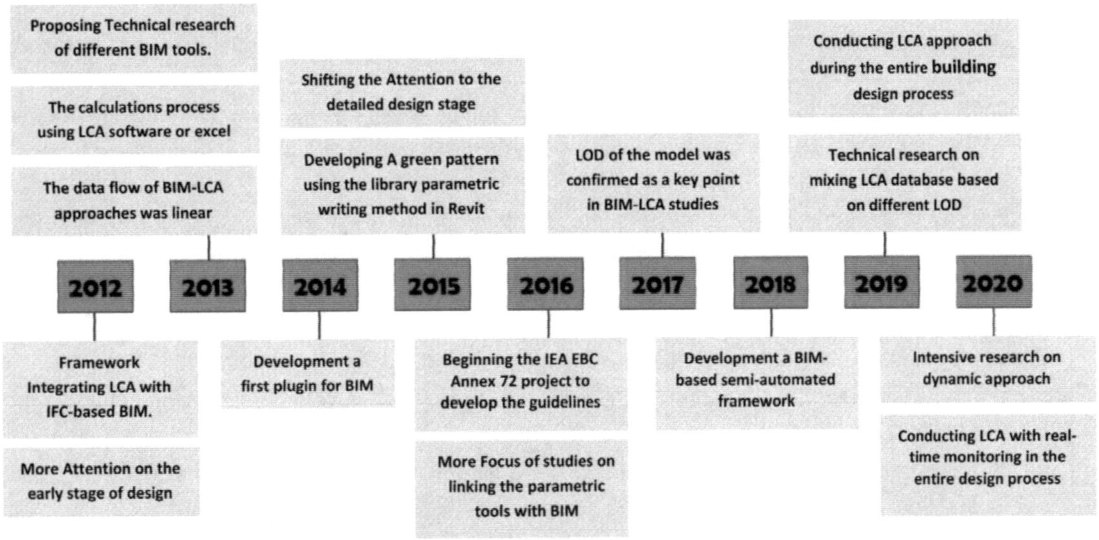

Fig 15: The BIM-LCA approach development since 2012. [28]

Concerning achieving the acceptable level of interoperability, there is rapid development in the integration criteria between the software. The initial attempts were in 2013 using the manual integration approach where the data are transferred by copying and pasting from excel to LCA software [29]. Then in 2014, a substantial trial to achieve a more seamless integration was began with developing the first LCA plug-in for the BIM tools. In 2018, a semi-automated approach was introduced to decrease manual data entries while the data are substituted and filled automatically and the user interaction is needed for the exports and imports. Finally, in 2020, a futuristic aspiration of new methodology; automated BIM-LCA integration, was launched where the data exchange is fully automatic. This futuristic aspiration is a promising approach for architects to have real-time monitoring and to provide instant feedback on the most essential impacts for design decisions. [30]

4. Conclusion

In order to figure out the past and future prospects, a comprehensive review was done. LCA is a recent discipline with 50 years of history and less than 30 years of intensive evolution in methodologies and applications. Nevertheless, throughout displaying the hotspot of LCA development in history focusing on the main five periods, there are massive efforts from all specialists in the field, policymakers, private sectors, and firms to develop the approaches, applications, international standardization, and broadcasting. To this day, LCA methodological development has continued to overcome the weak spot and guarantee that LCA will be the most widespread comprehensive tool for environmental evaluation.

It has become commonplace that the LCA traditional approach is a complicated process with a high potential of user mistakes owing to manual data extraction that must be handled and mapped. So, BIM-LCA integration is an innovative, promising methodology that introduces new prospects for especially architects and automates data extraction between the BIM and LCA software. Although, there are still various limitations of achieving the BIM-LCA interoperability which is the recent interest for all LCA specialists.

5. Reference

[1] J. Paulsen, Life Cycle Assessment for Building Products-The significanse of the usage phase, Institutionen for byggnader och installationer, 2002.

[2] N. Raynsford, "The UK's approach to sustainable development in construction," Building Research & Information, vol. 27, no. 6, pp. 419-423, 1999.

[3] S. Junnila, The environmental impact of an office building throughout its life cycle, Helsinki University of Technology, 2004.

[4] M. Najjar, K. Figueiredo, M. Palumbo and A. Haddad, "Integration of BIM and LCA: Evaluating the environmental impacts of building materials at an early stage of designing a typical office building," Journal of Building Engineering, vol. 14, pp. 115-126, 2017.

[5] M. M. Khasreen, P. F. Banfill and G. F. Menzies, "Life-cycle assessment and the environmental impact of buildings: a review," Sustainability, vol. 1, no. 3, pp. 674--701, 2009.

[6] G. Britain, Our energy future: creating a low carbon economy, Stationery Office, 2003.

[7] K. Simonen, Life cycle assessment, London: Routledge, 2014.

[8] M. Z. Hauschild, R. K. Rosenbaum and S. Olsen, Life Cycle Assessment: Theory and Practice, Springer, 2018.

[9] ISO, Environmental management: life cycle assessment; requirements and guidelines (ISO 14044), ISO Geneva, 2006.

[10] C. Bayer, M. Gamble, R. Gentry and S. Joshi, "AIA Guide to Building Life Cycle assessment in practice}," The American Institute of Architects, Washington DC, 2010.

[11] R. Rieckhof, "The life cycle metaphor: its emergence, understanding, and conceptualisation in business research," uwf UmweltWirtschaftsForum, vol. 25, no. 1, pp. 91-107, 2017.

[12] F. Bonviu, "The European economy: from a linear to a circular economy," Romanian J. Eur. Aff., vol. 14, p. 78, 2014.

[13] F. Sariatli, "Linear Economy versus Circular Economy: A comparative and analyzer study for Optimization of Economy for Sustainability," Visegrad Journal on Bioeconomy and Sustainable Development, vol. 6, no. 1, pp. 31-34, 2017.

[14] P. Lacy and J. Rutqvist, Waste to wealth: The circular economy advantage, Springer, 2016.

[15] A. Huhtala, "Circular economy: a commentary from the perspectives of the natural and social sciences," European Academies Science Advisory Council, Germany, 2015.

[16] N. Nazir, "Understanding Life Cycle Thinking and its Practical Application to Agri-Food System," International Journal on Advanced Science, Engineering and Information Technology, vol. 7, p. 1861, 10 2017.

[17] C. Dalhammar, "The application of 'life cycle thinking'in European environmental law: theory and practice," Journal for European Environmental \& Planning Law, vol. 12, no. 2, pp. 97-127, 2015.

[18] A. Azapagic, "Life cycle thinking and life cycle assessment (LCA)," Sustainable Development in Practice: Case Studies for Engineers and Scientists, p. 426, 2004.

[19] S. B. Young, Assessment of Environmental Life Cycle Approach for Industrial Materials and Products, Graduate Department of Metallurgy and Materials Science, University of Toronto, 1996.

[20] G. Benedetto, G. L. Corinto, B. Rugani, I. V'azquez-Rowe and E. Benetto, "The evolution of Life Cycle Assessment (LCA): some guidelines for future research in the wine sector," in Proceedings of the L Annual Meeting of SIDEA, Lecce, Lecce, 2013.

[21] . S. A. I. C. SAIC, Life-cycle assessment: principles and practice, Cincinnati: National Risk Management Research Laboratory, Office of Research, 2006.

[22] . J. B. Guinee, R. Heijungs, G. Huppes, A. Zamagni, P. Masoni, R. Buonamici, T. Ekvall and T. Rydberg, Life cycle assessment: past, present, and future, ACS Publications, 2011.

[23] F. M. Da Silva, H. M. Bartolo, P. Bartolo, R. Almendra, F. Roseta, H. A. Almeida and A. C. Lemos, Challenges for Technology Innovation: An Agenda for the Future: Proceedings of the International Conference on Sustainable Smart Manufacturing (S2M 2016), October 20-22, 2016, Lisbon, Portugal, Lisbon: CRC Press, 2017.

[24] J. Guinée, "Life cycle sustainability assessment: what is it and what are its challenges?," in Taking stock of industrial ecology, R. Clift and A. Druckman , Eds., UK, Springer, Cham, 2016, pp. 45-68.

[25] C. Wulf, J. Werker, C. Ball, P. Zapp and W. Kuckshinrichs, "Review of Sustainability Assessment Approaches Based on Life Cycles," Sustainability, vol. 11, no. 20, p. 5717, 2019.

[26] K. Forth, A. Braun and A. Borrmann, "BIM-integrated LCA - model analysis and implementation for practice," in IOP Conference Series: Earth and Environmental Science, vol. 323, IOP Publishing, 2019, p. 012100.

[27] C. Cavalliere, G. Habert, G. R. Dell'Osso and A. Hollberg, "Continuous BIM-based assessment of embodied environmental impacts throughout the design process," Journal of Cleaner Production, vol. 211, pp. 941-952, 2019.

[28] K. Safari and H. AzariJafari, "Challenges and opportunities for integrating BIM and LCA: methodological choices and framework development," Sustainable Cities and Society, no. 87, p. 102728, 2021.

[29] V. Durao, A. Costa, J. Silvestre, R. Mateus and J. De Brito, "Integration of environmental life cycle information in BIM objects according with the level of development," in IOP Conference Series: Earth and Environmental Science, 2019.

[30] J. P. Carvalho, L. Bragancca and R. Mateus, "Optimising building sustainability assessment using BIM," Automation in Construction, vol. 102, pp. 170-182, 2019.

[31] T. Ramesh, R. Prakash and K. Shukla, "Life cycle energy analysis of buildings: An overview," Energy and buildings, vol. 42, no. 10, pp. 1592-1600, 2010.

[32] A. Hollberg, G. Genova and G. Habert, "Evaluation of BIM-based LCA results for building design," Automation in Construction, vol. 109, p. 102972, 2020.

[33] J. Díaz and Á. A. Laura , "Sustainable construction approach through integration of LCA and BIM tools," in Computing in Civil and Building Engineering (2014), 2014, pp. 283-290.

IOP Conf. Series: Earth and Environmental Science **992** (2022) 012003 doi:10.1088/1755-1315/992/1/012003

A Study of Biophilic design and how it relates to the children's hospitals design

S G Abo Sabaa[1], M Abdel Azem[2] , H Al-Shanwany[3] and M El-Ibrashy[4]

[1]Architecture department , Faculty of Engineering , Egyptian Russian University, Egypt.

[2,3,4] Architecture department , Faculty of Engineering , Zagazig University , Egypt.

Saragamal674@yahoo.com

Abstract. As a modern design approach, biophilic design encourages the incorporation of natural components into the built environment, which has a substantial impact on human health, well-being, and productivity. The research aims to shed light on the architecture trends during the past period that addressed the design of children's hospitals, and among these recent trends is the biophilic design, which addresses with the impact of the natural environment surrounding users within health care buildings and its great impact on improving the mental health of children's hospital users, to suit it with the personality and tendencies of the child, and its reflection on raising the child's morale which brings him comfort during his treatment period, and increased recovery rates. The research is based on a common strategy between the historical and the descriptive approach, to evaluate the experience of adding biophilic design patterns as a competitive advantage to raise the efficiency of children's hospitals, which is reflected in providing a good experience for patients. The results will indicate Biophilic architecture is more inclusive than trends that uses or mimics nature in design, because of it is the genetic tendencies that humans have to be affected by nature, whether physical, emotional, or psychological.

Keywords: Biophilia, Biophilic architecture, Biophilic design, children's hospitals design, 14 Patterns of Biophilic Design

1. Introduction

Natural shapes drew many architects across generations as a source of fresh ideas that could be realized in design, and these forms were regarded as progressive visions of the future.
The interest in biophilic design, or the integration of natural components into spaces to create surroundings rich in positive experiences that can improve human health and well-being, peaked towards the turn of the third millennium.
Biologists and human medicine scientists in all their specializations possess extensive knowledge that can lead to a major breakthrough in the design of children's hospitals, but unfortunately in most biophilic design research, the differences between this trend and the trends that preceded it did not receive sufficient attention, and the researchers' interest was limited to clarifying the difference Between it and only one direction, which made a confusion between the different architectural trends, which prompted some designers to unleash without knowing the scientific and intellectual meaning of biophilic architecture.
Biology has attracted worldwide interest in the field of architecture where plants, animals, or entire biological systems are used as a basis for design, both because it is an important source of new innovations and because of the effects it has on creating a more healing environment.

2. Research Method and Scope

2.1. *Research problem*
First, Inadequate understanding and application of the concept of biophilic design, and limiting use in children's hospitals design. Second, Most of the designers mixed between architectural strategies related to nature on the one hand, and simulating nature on the other.

2.2. *Aims and objectives*
Clarify the differences between architectural trends associated with nature and biophilic design.

2.3. *Research hypotheses*
Biophilic architecture is more comprehensive than being a trend that uses or simulates nature indesign, but rather it is a genetic tendencies in humans to be affected by nature, whether that is physically, emotionally, intellectually, or even morally.

2.4. *Research questions*
Can biophilic architecture be employed to help development the design process in children's hospitals design? How do other strategies related to nature compare with the direction of biophilic architecture?

3. Literature Review

3.1. *Historical overview of (nature) in architecture*
The topic of how to conceptualize 'nature' as a concept is critical, because 'nature itself is not nature: it is concept, a standard, a recall, a good city, an alternative plan.' (Beck, 1999)
If we look at nature, we find that it has several meanings and connotations where it can be defined as the material world, including natural phenomena and living organisms, and it can also be viewed as the forces and processes that control natural phenomena in the material world independently of violations and interventions Human (Encarta, 2020).
Nature is a display of countless manifestations in life, to different degrees, whether living or non-living manifestations of (cells, organisms, waves, atoms...).
We find that the poet William Cowper described nature eloquently when he said that: (Nature is nothing but a name for the effect caused by God) (Cowper, 2002), and also described by the scientist, Ilya Prigogine (that life always informs us that there is something important in nature) (Prigogine, 1984). If we touch on the aspects of nature, we find that it consists of form and function. For the first aspect, we find that nature has provided us with countless forms, as the form is the first thing that is noticed by the eye, and it is also an embodiment of the materials used in nature. And the form as an attribute that characterizes things in every stage of tangible existence, and the aesthetic aspects that appear in an environment are a basic opportunity for the designer in particular and the artist in general to resort to as a rich dictionary of colors, lines, shapes, relationships and systems that link the elements in beautiful expressive formations, and the good artist is the one who He has the ability to contemplate nature and discern its beauty (Heerwagen, 2009).

3.2. *Defining biophilic design*
The term 'biophilia' was coined by social psychologist Erich Fromm to describe the 'love of life,' which indicated two major orientations for living beings: the first is to protect these organisms' lives from risks that cause death, and the second is to positively integrate with one other. (Fromm, 1964). It took 20 years for the biophilia theory to achieve widespread acceptance after it was initially introduced. 'The biophilia hypothesis,' according to biologist and naturalist Edward Wilson, is "the intrinsic predisposition to focus on life and life-like processes." (Wilson, 1984). Wilson also proposed the "biophilia hypothesis" to explain why people have an innate desire to focus on the processes that occur in life or in nature in general. This means that humans are completely reliant on nature and the resources and capabilities it contains to help and provide them with beauty, ideas, and knowledge, as

well as to spiritually enrich and satisfy them. (Wilson E. , 1993). He stressed that Biophilia is a study that studies the intrinsic emotional bond that exists between humans and biological organisms. Biophilia is related to nature and learning from it, just like other complex patterns of behavior that are governed by certain rules and regulations; thus, it cannot be replaced by any of the current man-made alternatives. (Kellert, 1993) Social ecologist Stephen Kellert explains the evolutionary dependency on "nature" by establishing nine values for biotopes: "utilitarian, natural, scientific, artistic, symbolic, humanistic, moral, sovereign, and passive." The last point could be a purposeful dilution of the term "innate," preventing biophilia from being confined to evolutionary psychology. (Joye, 2011)
Wilson, an American scientist, was able to define Biophilia as the intrinsic need to focus on life's or nature's processes in general. Kellert claims that biophilic design can promote positive interactions between people and nature, resulting in a "positive environmental effect."To put it another way, biophilic design is concerned with more than just lowering the environmental impact of the construction industry. While missing engagement with 'environment,' but building for Life: designing and understanding the human-nature connection (Kellert2005), This notion has gotten a lot of attention in the last two decades, thus there are a lot of theoretical research on biophilic design and its impact on human mental health, as well as its lack of use in buildings (Figure 1). (Berkebile, 2008)

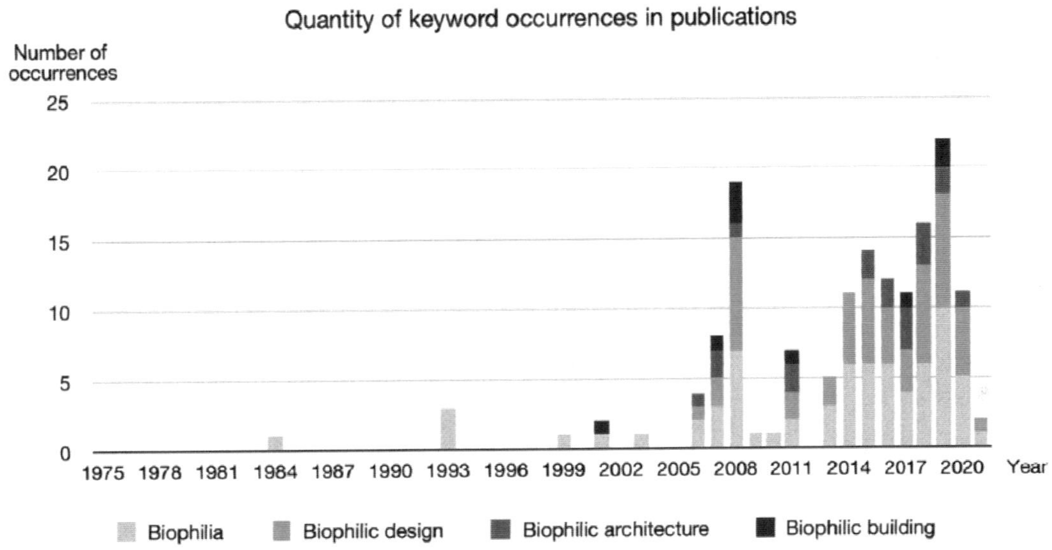

Figure 1. A figure showing in years the number of research studies in which biophilia and biophilic design have been studied in comparison with buildings that have already used biophilic design. (Weijie Zhong, 2021)

3.3. *Patterns to obtain Biophilia-based design*
Basic biophilic design patterns established by the environmental consultant Terrapin Bright Green are among the assessment items. They're called the "14 Patterns of Biophilic Design" because they show the connections between nature, human biology, psychology, and built environment design that can help designers implement and evaluate biophilic design successfully. (Browning, 2014)
The three categories of biophilic design - Nature in the Space, Natural Analogues, and Nature of the Space – provide a framework for understanding and intentional implementation of a diverse range of tactics into the architecture of children's hospitals. (Table 1)

3.3.1. Nature in the Space
Nature's direct, physical, and fleeting presence in a location is addressed in this area. Plant life, animals, water, noises, aromas, and other natural aspects can be included into the constructed environment to achieve this. (Browning W. R., 2020)

IOP Conf. Series: Earth and Environmental Science **992** (2022) 012003 doi:10.1088/1755-1315/992/1/012003

3.3.2. Natural Analogues

Organic, nonliving, and indirect evocations of nature" are what natural analogies are all about. These are frequently seen in the built environment's design, such as wood, furniture, decorations, and ornamentation, in objects, materials, colors, shapes, sequences, and patterns (Al-Rhodesly, 2019).

3.3.3. Nature of the Space

This category depicts natural spatial layouts, such as our fundamental urge to see beyond our immediate surroundings and to have visual exit options (Browning W. R., 2014).

Table 1 . The table shows The 14 Patterns of biophilic design (Browning W. R., 2020)

CONTEXT	PATTERNS
Nature in Space (7 patterns)	1. Visual Connection with Nature
	2. Non-Visual Connection with Nature
	3. Non-Rhythmic Sensory Stimuli
	4. Thermal and Airflow Variability
	5. Presence of Water
	6. Dynamic and Diffuse Light
	7. Connection with Natural systems
Natural Analogues (3 patterns)	8. Biomorphic Forms and Patterns
	9. Material Connection with Nature
	10. Complexity and Order
Nature of The Space (4 patterns)	11. Prospect
	12. Refuge
	13. Mystery
	14. Risk/Peril

3.4. *Architectural trends associated with nature:*

Over the years, many architectural trends have emerged, which some agree as emanating from the organic school, with the aim of ridding contemporary architecture of monotony and rigidity, even if it sometimes tends to exaggerate its forms in the natural world. They also see in it an invitation to give adequate attention to the sensual and spiritual aspect of man. This is an important aspect that has been neglected by contemporary architecture in its successive quest to meet the material and utilitarian needs of man. It can be summarized in four main directions: environmental architecture, green architecture, sustainable architecture, and bio-architecture (Figure 2). (Capra, 2002)

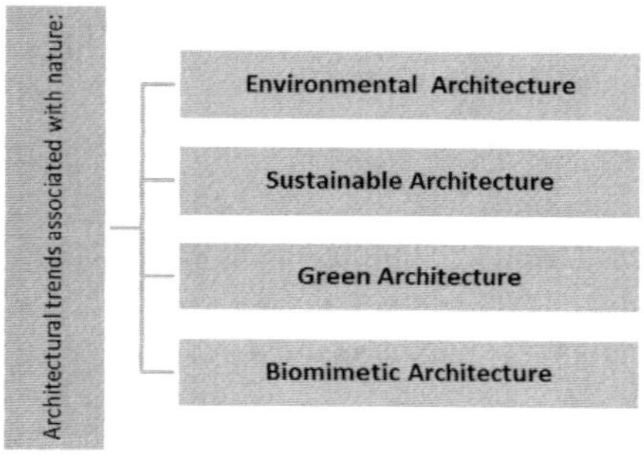

Figure 2. A figure illustrating the architectural trends associated with nature that is confused with biophilic architecture (Source: authors)

3.4.1. Environmental architecture

Environmental architecture appeared since antiquity, then civilizations followed and humans dispensed with this thought relatively as a result of progress and the search for luxury, and the need to look at the concepts of environmental architecture to lessen the negative environmental impact, began in the seventies the emergence of some steps towards this direction and followed in the nineties More

advanced to shed light on human consumption of the natural resources of the earth and reduce pollution rates, and many architectural schools appeared that called for the building to be part of the environment and to preserve it and not to disturb it. Sustainable architecture, etc ***(Abdullah, 2012)***.
Environmental architecture is defined as "architecture that considers environmental conditions in the design of buildings and the selection of materials in order to achieve thermal comfort for its users; to building compatibility with the surrounding environment and access to an architectural design to suit the climatic diagnosis and human environmental needs," according to some. (Siraj, 1990)

3.4.2. Sustainable Architecture

It's a broad phrase that refers to ecologically responsible architectural design strategies. It is the process of designing buildings in a way that is environmentally friendly, taking into account minimizing energy, material, and resource consumption as well as reducing the consequences of construction and use on the environment while maintaining natural harmony. Then, by examining the pressing economic and political concerns in our world, clarify sustainable architecture from a broader perspective. Sustainable design, on a large scale, aims to limit buildings' negative environmental impacts by maximizing the effective use of materials, energy, and space. (Sustainable Architecture and Simulation Modeling", 2018)

3.4.3. Green Architecture:

Green architecture, or environmentally friendly buildings and cities, is one of the modern trends in architectural thought, which is concerned with the relationship between buildings and the environment, and there are many concepts and definitions that have been developed in this field.
Green architecture, according to architect Ken Yang, must meet the needs of the present without ignoring the right of future generations to meet their own needs, and green buildings, according to architect William Reed, are buildings that are designed, implemented, and managed in a way that considers the environment, and He also sees that one of the concerns of green buildings is reducing the building's impact on the environment. (Akhimien .N, 2017)
Green design is one of the modern trends in architectural thought, which is concerned with the relationship between buildings and the environment, and green architecture is a highly efficient system that matches its biosphere with minimal damage to the environment. The environment and its sources, hence the description of this architecture as green as plants, as it achieves success and adaptation in its place as it takes full advantage of the ocean to obtain its nutritional requirements (Muhammad Arshid, 2015)

3.4.4. Biomimetic architecture

The term design that mimics nature was defined in the twenty-first century, in which the world called for a shift towards nature and attention to the surrounding environment, although looking at nature is not new, but the angles of visions differ from one period to another: such as the appearance of paintings and drawings containing plant or animal decorations on the walls of the temples of ancient civilizations And the use of human and animal statues in palaces and entrances to important public buildings (Benyus, 1997).
The process of simulating nature is a two-way process: the first begins with conducting many biological and scientific research on certain characteristics or behaviors in an organism or an ecosystem, then determines areas and possibilities of exploiting and applying them in the form of design ideas, and the second begins with identifying a need or a design problem, then an attempt Searching for solutions to them by observing living organisms and ecosystems, with the help of previous biological research and available information, and in the latter case the designer often lacks in-depth scientific understanding in his access to the possibility of simulating nature. These trends can be defined as those that examine natural models and then draw inspiration from designs and processes to solve human problems, and it is a broad and inclusive term for many notions that are synonymous with design emulating nature. (Dosen, 2013)

Biomimetic architecture is a modern architectural concept that seeks solutions for sustainability in nature through knowledge of the rules that govern those forms, rather than by the reproduction of natural shapes. It's an interdisciplinary approach to sustainable design that adheres to a set of principles rather than a set of stylistic guidelines. It's part of a bigger movement called biological movement, which examines nature and its models, systems, and processes in order to find inspiration for fixing man-made problems. Biomimetic design not only aids in the discovery of new and sustainable architectural solutions, but it may also be used in other ways to meet human needs. (Youssef, 2021)

4. Materials and Methods

The research initially presented a historical overview of (nature) in architecture to study the characteristics of Biophilic design for children's hospitals. Defining biophilic design and summarizing the elements that were included in biophilic design patterns in previous studies. Second, we analyzed the Architectural trends associated with nature; environmental architecture, green architecture, sustainable architecture, and Biomimetic architecture. The article will build and revise a definition that is based on the literature and addresses the study's conceptual and theoretical assumptions. The extent to which biophilic design is related to (architectural trends associated with nature), as well as the extent to which biophilic design patterns incorporate the fundamental principles on which these trends are built, are discussed in this study.

In the following table (

) are put in vertical direction biophilic design patterns developed by environmental consultancy Terrapin Bright Green. It is the "14 Patterns of Biophilic Design" to show the relationships between nature, human biology, psychology, and the design of the built environment that can help designers in the successful implementation and evaluation of biophilic design, and in the horizontal direction are put architectural trends associated with nature; Environmental architecture, green architecture, sustainable architecture, and biomimetic architecture, and the basic principles on which each direction is based were developed to measure their compatibility with biophilic design patterns, and then the points obtained for each direction were calculated to find out which directions are closer to the biophilic design .

From this table, it was concluded that architecture Biomimetic architecture is the closest approach to biophilic design, then environmental architecture, green architecture, and finally sustainable architecture (Figure 3).

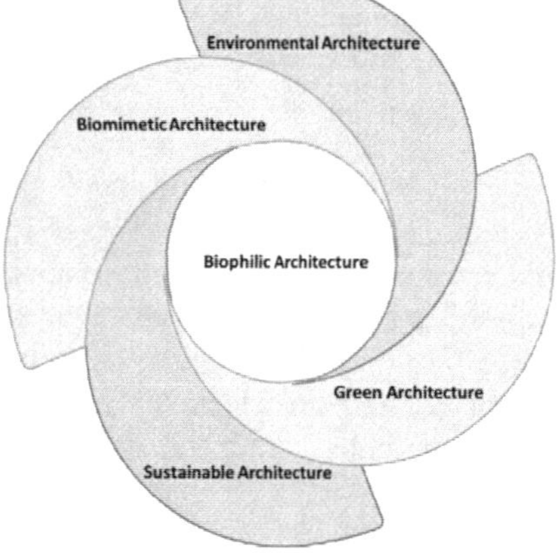

Figure 3. The research can conclude that biophilic architecture is a mixture of architectural trends related to nature (Source: authors)

IOP Conf. Series: Earth and Environmental Science **992** (2022) 012003 doi:10.1088/1755-1315/992/1/012003

Table 2. Comparison of architectural trends related to nature with biophilic styles (Source: Authors)

Architectural trends associated with nature	biophilic design patterns	1.Visual Connection with Nature	2.Non-Visual Connection with Nature	3.Non-Rhythmic Sensory Stimuli	4.Thermal & Airflow Variability	5.Presence of Water	6.Dynamic & Diffuse Light	7.Connection with Natural Systems	8.Biomorphic Forms & Patterns	9.Material Connection with Nature	10. Complexity & Order	11.Prospect	12.Refuge	13.Mystery	14. Risk/Peril	All points	Total
Biomimetic architecture	Creativity in training is the result of adaptation, benefit and efficiency	●	●	●	●	●	●	●	●	●	●	●				11	44
	Examine nature, its models, systems, and processes for inspiration Maximize resources by rationalizing inputs and rotating outputs	●	●	●	●	●	●	●	●	●	●	●	●	●	●	14	
	The shapes are derived from the environmental environment and have the ability to grow and expand in the future	●	●	●	●	●	●	●	●	●	●	●	●	●	●	14	
	Form follows function, adapts to the environment, and is shaped by matter	●						●	●	●		●				5	
Green Architecture	Providing a healthy environment	●	●	●	●	●	●	●	●	●	●	●	●	●	●	14	36
	water conservation															0	
	Preserving the environment and the nature surrounding the building	●	●	●	●	●	●	●		●		●				9	
	Reducing the use of building materials and recycling building materials									●						1	
	Reduce environmental pollution, waste					●		●		●						3	
	Energy conservation and green energy production	●			●		●	●		●						5	
	Reducing energy consumption from non-renewable energy sources				●		●	●		●						4	
Sustainable Architecture	Efficient use of resources	●			●	●	●			●		●				6	15
	Harmony with the environment (including environmental assessment)	●								●		●				3	
	Using materials that have a benign effect on the environment, preventing pollution					●		●		●						3	
	Achieving energy efficiency				●		●			●						3	
Environmental Architecture	The acceptable limit of healthy conditions for human living	●	●	●	●	●	●	●	●	●	●	●	●	●	●	14	37
	Achieving a minimum level of environmental pollution.	●								●		●				3	
	Preserving the available environmental resources, maximizing harmony with nature.	●	●	●	●	●	●	●	●	●		●				10	
	Reducing the impact of construction and use on the environment.				●		●	●		●		●				5	
	The compatibility between economic performance and social responsibility.							●								1	
	Respecting the environmental dimension in architectural work.	●			●			●		●						4	

5. The Selected Case Study of the Biophilic Children's Hospital: The Royal Children's Hospital

At the prestigious World Architecture Festival (WAF) Awards in Singapore, the Royal Children's Hospital in Melbourne earned the 'World Health Building of the Year 2012' Award. A jury of some of the world's most influential architects and urban planners chose the structure. The hospital fostered the idea of patient-environment harmony through well-integrated design with nature, such as picking the building's site, furniture design, and broad windows, as well as making the hospital a part of the surroundings. (Batessmart, 2012)

Table 3. Information about the Royal Children's Hospital (The Royal Children's Hospital Melbourne, 2020)

Information about the hospital	
Location:	50 Flemington Road, Parkville Victoria 3052, Melbourne, Australia
Architects:	Billard Leece Partnership and Bates Smart
Building area	165000 m^2
Beds:	340
Completion date:	2011

The Many parts of the hospital's design were inspired by the nearby Royal Park's original bush land environments. Courtyards, a sweep of colored 'leaves,' panoramic vistas of the parkland, a two-story coral reef aquarium, large-scale artworks, and a miniature zoo were among the innovative architectural elements employed to optimize the connection to nature. Surprisingly, these new design ideas have significantly reduced the hospital's users' pain, worry, and anxiety by normalizing tactile and engaging environmental experiences (Figure 4). (Detail. Nature-Inspired Design, 2013).

Figure 4. The Royal Children's Hospital (RCH) is a major children's hospital in Melbourne, Australia. (The Royal Children's Hospital Melbourne, 2020)

Table 4 . Applications of Biophilic Design Patterns in Royal Children's Hospital (Source: authors)

Biophilic design patterns		Applications of Characteristics of Biophilic Design Elements and Patterns in Royal Children's Hospital
Group one: Nature in Space (7 patterns)	1-Visual Connection with Nature	 This pattern was achieved by looking at the elements of nature, living systems and natural processes, and it was easily applied in this hospital due to the presence of the Royal Park around the hospital from three sides, Space design that can stably and comfort ably allow for gazing at the nature outside, while seated

	indoors in Outpatient waiting rooms and patient rooms.
2-Non-Visual Connection with Nature	 This pattern was achieved by auditory, tactile, olfactory, or gustatory stimuli that generate a deliberate, positive cue to nature, living systems, or natural processes. Provision of natural tactile and olfactory stimulation through outside recording space that has functions such as gathering, play, and performance.
3-Non-Rhythmic Sensory Stimuli	 This pattern was achieved by Stochastic and transitory interactions with nature that can be statistically examined but not accurately anticipated.
4-Thermal & Airflow Variability	This pattern was achieved by Natural-looking variations in air temperature, relative humidity, airflow across the skin, and surface temperatures which Positively impacted comfort, well-being and productivity; The patient's room provides a feeling of both flexibility and a sense of control.
5-Presence of Water	 A condition that improves a person's perception of a place by allowing them to see, hear, or touch water. This pattern was achieved by the fountains scattered around the hospital and in the children's play areas, as well a two-story coral reef aquarium which are located in the waiting room.
6-Dynamic & Diffuse Light	 Using different intensities of light and shadow that change over time to produce natural-looking settings, Exposing the sky through the ceiling, providing a natural ecosystem view and dynamic light in the environment, and in the moving space.
7-Connection with Natural Systems	 This pattern is achieved by Natural processes, particularly seasonal and temporal fluctuations are important aspects of a healthy ecosystem. Through the outpatient courtyard features two-toned synthetic turf and concrete donuts, there is a space structure where visitors can feel the changes of nature, such as seasons and weather, and moving spaces. Meerkats have been quarantined inside the Royal Children's Hospital. The objective of the meerkat display is to entertain children and divert their attention away from the fact that they are in hospital, which may be distressing to them, but the meerkat exhibit will make

		them happy and assuage their concerns. And also the two-storey reef aquarium is visible from Emergency on Lower Ground and Main Street on Ground floor .It is home to approximately 530 fish of 32 different species. All species can be found on the Great Barrier Reef, Australia.
Natural Analogues	8-Biomorphic Forms & Patterns	 It is composed of paintings and sculptures symbolizing the nature and shapes of nature, such as trees and forests.
	9- Complexity & Order	Provision of an environment where patients can receive various sensory information from nature in natural center spaces, such as landscaping, green walkways, and water.
Nature of The Space	10- Prospect	 The hospital is surrounded by the parkland of Royal Park, with views of trees and much natural light, The unrestricted views of the Hospital extend prospect with sightlines down to the lower parts of the complex, convey the feeling of openness and freeing.

6. Results and discussion

The debate continues among designers about the best strategies for designing hospitals, especially children's hospitals, to increase recovery rates, and thus reduce the length of stay for the patient. This study discussed some of these strategies, especially those related to nature, to find out which is better for children's hospitals, and the results were as follows:

- Biophilic architecture is more inclusive than trends that uses or mimics nature in design, because of it is the genetic tendencies that humans have to be affected by nature, whether physical, emotional, or psychological.
- Biophilic design is a mixture of the following architectural trends: environmental architecture, green architecture, sustainable architecture, and biomimetic architecture. When comparing these trends and the 14 patterns of biophilic design, it was found that most of the patterns include the basic principles of these trends.
- Biomimetic design is closer to biophilic design with a number of points (44), which is the largest number, followed by environmental design with (37), green design with (36) and finally sustainable design with (15).
- Biophilic design is one of the trends that care about human health in the first place, which makes it the most suitable for application in children's hospitals.

7. Conclusion

This study delves into the notion of biophilic design, including how it came to be, how it is defined, how it may be used to children's hospitals architecture, and what the key design strategies are. We started by looking into the underlying theories and interpretations of biophilic design. The numerous benefits of biophilic design and the multiple problems in architectural trends associated with nature were then compared and integrated. We proposed a biophilic design framework and assembled design methods, bringing together the fundamental principles of many natural aspects in building.

8. Recommendations

Many uncertainties remain, given that biophilic design has only been used in building for 20 years. Although the concept of biophilic design gives many design inspirations, architectural terminology is rarely employed to describe children's hospitals. Future study should look at examples from an architectural standpoint, taking form and technology into account. According to the findings, the following actions should be taken: The importance of a relationship and communication between architectural and interior architecture, on the one hand, and nature, on the other. Increasing architects' understanding of the relevance of environmental design, current and modern trends, and natural simulation. Ensure the application of biophilic design foundations in the fields of architecture and interior architecture. Supporting research and specialized study in areas related to environmental preservation.

9. References

[1] *ADR .Royal Children's Hospital*. (2012, May 14). Retrieved Nov 1, 2021, from Australian Design Review: https://www.australiandesignreview.com/architecture/royal-children'shospital/
[2] *Batessmart*. (2012). Retrieved Oct 2021, from https://www.batessmart.com/2012/sep-oct/royal-childrens-hospital-melbourne/ .
[3] *Detail. Nature-Inspired Design*. (2013, March 15). Retrieved Nov 22, 2021, from The Royal Children's Hospital in Melbourne: https://www.detail-online.com/article/natureinspired-
[4] *Sustainable Architecture and Simulation Modeling"*. (2018). Dublin Institute of Technology.
[5] *Arch.Ahlamontada*. (2020). Retrieved 2021, from https://arch.ahlamontada.net/t250-topic.
[6] *The Royal Children's Hospital Melbourne*. (2020). Retrieved 2021, from https://www.rch.org.au/home/.
[7] Abdullah. (2012). *"A Sustainability Approach, Towards a Traditional Square Design Model"*. Faculty of Engineering, Misr International University, Ph.D. thesis, p.3,4.
[8] Akhimien .N, I. A. (2017). *Green Architecture as a panacea to the Environmental effects of Architectural development*. Department of Architecture, Ambrose Alli University, Ekpoma, Nigeria.
[9] Al-Rhodesly, A. M. (2019). *Biophilic Approach to Architecture:Case of the Alhambra, al-Andalus*. Architectural Engineering Department , Faculty of Engineering, Alexandria University.
[10] Beck, U. (1999). *World risk society as cosmopolitan society? Ecological questions in a framework of manufactured uncertainties.* . Cambridge: World Risk Society. Polity Press,pp. 19-23.
[11] Benyus, J. (1997). *Biomimicry - Innovation Inspired by Nature*. New York, Harper Collins Publishers.
[12] Berkebile, B. F. (2008). *Reflections on implementing biophilic design. In: Biophilic Design: the Theory, Science and Practice of Bringing Buildings to Life*. pp. 347-356.
[13] Browning, W. R. (2014). *14 Patterns of Biophilic :Improving Health & Well-Being in the Built Environment*. New York: Terrapin Bright Green,LLC.
[14] Browning, W. R. (2020). *What is biophilia and what doesit mean for buildings and spaces? Nature Inside: A Biophilic Design Guide*. RIBA Publishing, pp. 1e5.
[15] Capra, F. (2002). *The Hidden Connection : Integrating the Biological, Cognitive, and Social Dimensions of life into a Science of Sustainability*. Double day, NewYork, U.S.A, P14.
[16] Cowper, W. (2002). *The Task, book VI, The Winter Walks at Noon, in Ahmed Yehia, Mostafa: Architecture as a manifestation of mans understanding of nature*. Cairo: Masters of science in Architecture, Cairo University , P.12.
[17] Dosen, A. O. (2013). *Prospect and refuge theory: constructing a critical definition for architecture and design.* . Int. J. Des. Soc. 6, 9- 23.
[18] Encarta. (2020). *World English Dictionary [North American Edition]*. Microsoft corporation by Bloomsbury publishing plc.
[19] Fromm, E. (1964). *Love of death and love of life. In: The Heart of Man: its Genius for Good and Evil*. New York: New York: Harper & Row.

[20] Heerwagen, J. (2009). *Biophilia, health, and well-being.* USA: Pennsylvania, USA: USDA Forest Service, Northern.

[21] Joye, Y. d. (2011). *"Nature and I are two": a critical examination of the biophilia hypothesis.* . Environ. Val. 20, 189- 215.

[22] Kellert, S. (1993). *The biological basis for human values of natureIn: Kellert, S.R., Wilson, E.O. (Eds.), The Biophilia Hypothesis.* Island Press, pp. 42-69.

[23] Kellert, S. (2005). *Building for Life: Designing and Understanding the Human-Nature Connection.* Island Press, Washington.

[24] Muhammad Arshid, A. (2015). *Applications of green architecture in the interior design of Irbid heritage houses.* Master's thesis, University of Islamic Sciences, Jordan, page 9.

[25] Prigogine, I. (1984). *Order out of Chaos, Bantam New York. In Capra, Fritjof,(1996).* The Web of Life, NY:Anchor Book, P.86.

[26] Siraj, N. M. (1990). *Energy-Guiding Architectural Design in Radiation Research Buildings.* Faculty of Fine Arts (Department of Interior Architecture), Helwan University, PhD thesis in Environmental Sciences, pp. 54, 55.

[27] W. Zhong, T. S. (2021). *Biophilic design in architecture and its contributions to health, well-being, and sustainability: A critical review.* Netherlands: Architectural Design and Engineering, Department of the Built Environment, Eindhoven University.

[28] Wilson, E. (1984). *Biophilia: The Human Bond with Other Species.* MA: Harvard University Press, Cambridge.

[29] Wilson, E. (1993). *Biophilia and the conservation ethic. In: Kellert, S.R., Wilson, Edward Osborne (Eds.), The BiophiliaHypothesis.* Island Press., P.20- 31- 32.

[30] Youssef, Y. R.-E. (2021). *Biomimetic approaches to sustainability and its application in interior design of the tourist*. International Design Journal, Volume 4, Issue 2.

IOP Conf. Series: Earth and Environmental Science **992** (2022) 012004 doi:10.1088/1755-1315/992/1/012004

Studying the influence of shading devices on indoor thermal comfort in desert and Mediterranean climates

D Berkouk[1], TAK Bouzir[2], S Mazouz[3], S Boucherit[4,*] and N Mokhtari[4]

1 Department of Architecture, Biskra University, Biskra 07000, Algeria
2 Institute of Architecture and Urban Planning, Blida University, Blida 09000, Algeria
3 LEQUAREB Laboratory, Department of Architecture, Oum El Bouaghi University, Oum El Bouaghi 04000, Algeria
4 Department of Architecture and Industrial Design, Università degli Studi della Campania "Luigi Vanvitelli", 81031 Aversa (CE), Italy
E-mail: samiha.boucherit@unicampania.it

Abstract. Thermal comfort is regarded as the most important issue in residential buildings. To verify the thermal performance of the envelope components used for the promotional apartments in Algeria, whether they are adapted to the hot and dry southern climate, or the Mediterranean climate, and to understand the effect of fixed and combined horizontal and vertical shading devices of different dimensions (12 cm, 25 cm, and 50 cm) on the indoor thermal comfort of these apartments, two main climatic zones have been considered in this study, the first is a desert climate of (Biskra, Ghardaïa, and Ouargla), while the second is a Mediterranean climate of (Algiers, Oran, Bari, and Naples). The methodology of this research was based on a comparative approach to the outcomes of a stay in an apartment in Biskra, which was looked at as a model for other desert and Mediterranean cities, based on the operative temperature data of different scenarios obtained through several simulations, which conducted by the use TrnSys software. The findings suggest that the double-walled hollow brick envelope causes a very severe indoor thermal environment (To<29) in dwellings located in desert regions such as Biskra, Ghardaïa, and Ouargla. In addition, the findings also suggest that shading devices are more effective for apartments located in the Mediterranean region (Algiers, Oran, Bari, Naples) than those located in the desert region. In the desert region, the use of mechanical cooling devices in the summer period is considered the most adequate solution to ensure the occupants' thermal comfort. To identify the most optimal shading device models, future research based on the parametric simulation of kinetic shading devices is needed.

Keywords: Fixed shading devices; Indoor thermal comfort; Numerical simulation; Promotional apartment; Climate.

1. Introduction

Thermal comfort is considered one of the essential parameters in the building design process [1], where it can be a key issue in building performance [2,3] because the indoor thermal conditions are rarely stable due to the interaction between the building envelope [4], climate and occupancy [5]. This is the reason why the research on optimal indoor thermal conditions of buildings has become a priority necessity, and this is to ensure the comfort of occupants and to minimize the energy consumption of the building at the same time [6,7]. In the scientific literature, the theories of thermal comfort in workplaces during the day are widely studied [8], and even in residential environments, however, few researches have been carried out to assess the thermal comfort of residential buildings in the North African countries.

In Algeria, there are two main regions that are characterized by two different climates which are the northern region characterized by a Mediterranean climate and the southern one with a desert climate [9]. The building sector in this country consumes about 43% of total energy consumption [10], which is similar to the building energy consumption of the European Union, and the United States (40%) [11], therefore, many researchers believe that the building construction sector is the main source of greenhouse gas emissions [12]. In Algeria, the building sector consumes the largest part of energy compared to that consumed by the transport sector 36%, and industry 21% [13]. Moreover, residential buildings in the southern regions consume approximately 75-80% of the total electrical energy during the hot period, from the beginning of May to the end of September [14], because the building policy in southern Algeria is similar to that of the northern ones [9] despite having different climates. This building policy causes serious economic problems in Algeria, which has been obliged to follow an austerity policy since the beginning of 2016, due to the decline in oil prices. Therefore, this situation should not be overlooked.

The residential buildings in Algeria have become energy-intensive, because the occupant is constantly to ensure or improve the comfort of his apartment by using non-renewable energy sources, where he resorts to heating in the winter, and the air conditioner in the summer. That is why the main priorities in the design and operation of buildings must be human comfort and energy efficiency [15], where the various components of the building envelope (walls, windows, and roof) must be the key to minimizing energy demand and to ensure the indoor thermal comfort at once [16]. For this, architects must understand and control the interaction between these architectural components, and the environment (indoor and outdoor), from this, the climate is considered as an imperative factor in determining design parameters such as building orientation and building envelope [17]. This opens the need to assess the indoor thermal performance of residential buildings in Algeria to invite decision-makers to review the construction laws that require architects to use the same building materials for the different climatic zones of this country, considering the climate as a requirement, and passive design techniques as a solution [18], to improve the indoor comfort level of residential buildings, which subsequently minimizes the energy consumption of this sector.

The main objective of this research is to understand the effect of shading devices on the indoor thermal comfort of apartments in different climatic zones, such as Biskra, Ghardaïa and Ouargla (Algerian desert climate), Algiers and Oran (Algerian Mediterranean climate), and Bari and Naples (Italian Mediterranean climate). It also aims to verify the thermal performance of the envelope components used in the construction of apartments in Algeria, if they are more adapted to the hot and dry climate of the south, the Mediterranean climate of northern Algeria, or the Mediterranean climate of southern Italy.

2. Methodology

This research is based on a comparative approach between different scenarios of shading devices size and climate of the apartment in different orientations such as north, south, east, and west orientation, based on a quantitative evaluation of indoor thermal comfort. For this, the collective promotional apartment [19] was considered as a case study, where the choice was on a typical living room of an apartment, oriented to the four cardinal directions, of the city 110 apartments that are located in the city of Biskra, Figure 1. The constructive system of these residences is a "post-beam» structure, of a height of 2.9 m for each floor. In addition, five cities were also considered with two different climates (desert and Mediterranean climate) to study the effect of climate on the indoor comfort level of this type of housing, and to verify the thermal performance of the components of the envelope of the promotional apartments in the hot and dry climate as those in the city of Biskra [20], where the cities of Ghardaïa and Ouargla were chosen for the Algerian desert climate, Algiers and Oran for the

Algerian Mediterranean climate, while Bari and Naples were obtained for the Italian Mediterranean climate (See Figure 1).

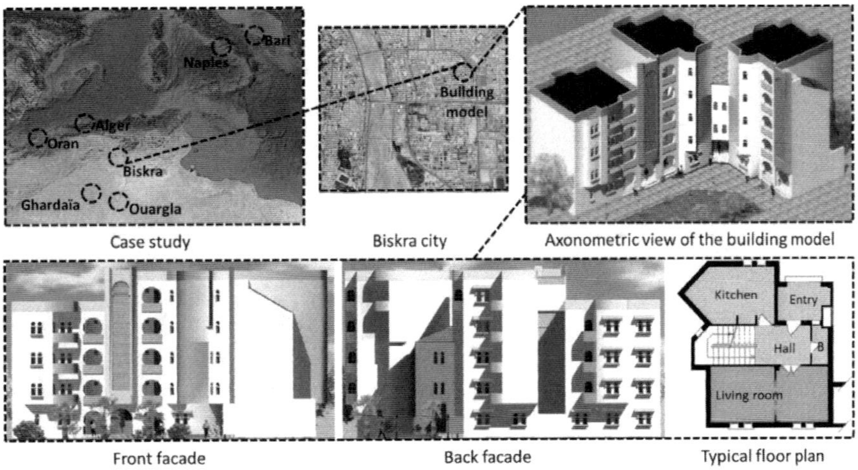

Figure 1. Presentation of case studies.

In this study, some shading device scenarios used in the study of Ali Ahmed [17,21], and Berkouk et al. [22,23] were combined to verify and confirm the results of these previous studies. The different scenarios of combined shading devices (vertical and horizontal) were; an apartment with no shading device (0 cm), with a device of (12 cm), and of (25 cm), and an apartment with a shading device of (50 cm) size. Furthermore, the acceptable thermal sensation of architectural spaces occupants is greatly affected by environmental factors [21], as these factors have a direct impact on the physiological state of the human being in architectural spaces [24]. Regarding thermal comfort, the temperature is considered the most important environmental parameter [5], so the indicator that was considered in this research was the operative temperature, which is expressed as the relationship between air temperature and radiant temperature, including the thermal exchanges by convection and radiation between the occupant and its environment [25], and ignoring the effect of humidity and air movements. In the scientific literature, numerical simulation is thought of among the most advanced tools for assessing thermal comfort, because it can be applied to buildings in different climates [24]. For this reason, the numerical simulation based on the study of the operative temperature as a tool for this study to evaluate the indoor thermal comfort of apartments through the use of TRNSYS software as the main research tool.

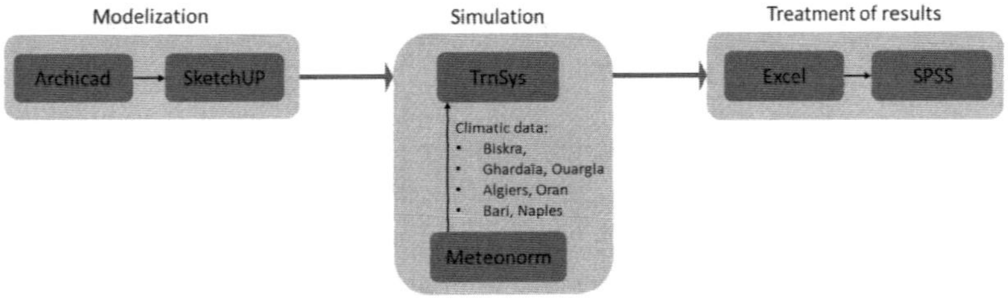

Figure 2. Working process.

The process of this work is demonstrated in Figure 2. The modeling was performed by ArchiCAD (for 2D) and Google SketchUp software (for 3D). The three-dimensional models were defined and subdivided into thermal zones by applying TrnSys plug-ins so that the volume was exported to TRNSYS studio software. It is noted that the same scenario of internal inputs was considered in TRNBuild and Multizone Building modeling with Type 56 for the different scenarios. It is also interesting to note that the same occupancy scenario was also used for the different models, where this study was based on a five-person occupancy scenario for each apartment. Hence, the infiltration, the occupants' metabolic energy and the thermal resistance of the clothes were fixed, respectively, at 0.8 vol/h, 1.2 met, and 1.5 clo. To avoid the air currents effect, the window closure was considered as the scenario of the window condition. The same internal gains of equipment were obtained, such as a TV of 540 KJ/h and a computer, while the lighting power was set by four lamps.

Table 1. Envelope materials and their thermal characteristics [26].

Walls / Floors	Materials	Thermal conductivity λ [W/m.°C]	Thickness [m]
Exterior wall	Cement plaster	1,4	0,015
	Hollow brick	0,5	0,15
	Air blade	0,31	0,05
	Hollow brick	0,5	0,1
	Gypsum plaster	0,35	0,015
Interior wall	Gypsum plaster	0,35	0,015
	Hollow brick	0,5	0,1
	Gypsum plaster	0,35	0,015
Low and intermediate floors	Gypsum plaster	0,35	0,015
	Hollow body + compression slab	1,45	0,2
	Mortar	1,4	0,04
	Flooring	2,1	0,06
Terrace floor	Gypsum plaster	0,35	0,015
	Hollow body + compression slab	1,45	0,2
	Isolation	0,1	0,04
	Slope shape	1,15	0,04
	Watertightness	0,04	0,03

Regarding the integration of building materials, the external walls were considered double walls of hollow bricks (see Table 1). This constrictive system of walls consists of a first wall of bricks of 15 cm towards the outside, and another of bricks of 10 cm to the inside, with a separation of a 5 cm air gap. They are coated with cement plaster on the external side and plaster with a thickness of 1.5 cm on the internal side. In addition, the interior walls are considered simple brick walls of 10 cm, while the ceiling and planks were hollow slabs. In addition, the integration of weather data of the different selected cities (Biskra, Ghardaïa, Ouargla, Algiers, Oran, Bari, Naples) which were exported from the Meteonorm 7 software in TMY2 format was done using the TRNSYS type 109 software. Finally, the statistical study was carried out mainly by SPSS software "Statistical Package for the Social Sciences".

3. Results

3.1. Comparison of the thermal environment of different scenarios of housing with no shading devices

To verify whether the promotional housing is suitable or not in the desert climate like that of the city of Biskra, several simulations of the operative temperature have been made in Biskra, Ghardaïa, and Ouargla to take an idea on the thermal performance of the constructive system of the double wall in hollow brick used in these apartments in the desert climate of southern Algeria, and simulations were also made in Algiers, Oran, Bari, and Naples to check the effectiveness of this system in the north of the country and the Mediterranean climate in general. Table 2 shows the cumulative percentages of operative temperature hours of different desert and Mediterranean climate scenarios of different housing without shading devices. While Figure 3 shows the hour's percentages of three comfort zones according to the operative temperature of various scenarios of these apartments.

From Figure 3 and Table 2, it is observed that the values obtained from the simulated housing in cities that are characterized by a desert climate (Biskra, Ghardaïa, and Ouargla) are included between the two discomfort zones due to the heat: the discomfort zone (26<To<29), and the very discomfort zone defined as thermally aggressive (To<29), concerning the various orientations (North, South, East, and West). Where the hours' percentages of the thermally aggressive zone are higher than those of the discomfort zone. The hours' percentages of the very discomfort zone are between 91.75% in the south, and 97.87% in the east, concerning the housing of the city of Biskra, while those located in Ghardaïa are characterized by hours' percentages vary between 79.05% in the south, and 88.72% in the east, while the percentages of hours that vary between 99.95% in the South and 97.37% in the East characterize the housing located in Ouargla.

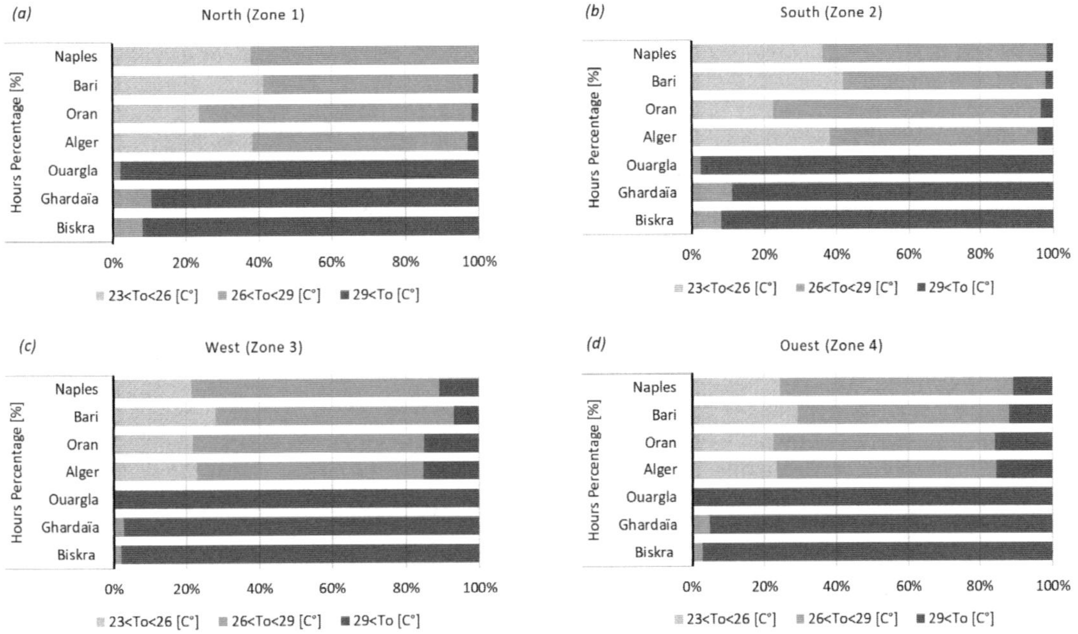

Figure 3. Percentage of comfort zone hours according to the operating temperature of different climate scenarios of various apartments without shading devices (0 cm).

Table 2. Cumulative percentage of hours of the operative temperature of different climate scenarios of various apartments without shading devices.

Orientations	Operative temperature	Cumulative percentage [%]						
		Biskra	Ghardaïa	Ouargla	Alger	Oran	Bari	Naples
North	23 [C°]	0,00	0,00	0,00	10,15	22,05	17,89	10,33
	26 [C°]	0,00	0,00	0,00	44,71	40,62	51,91	44,43
	29 [C°]	8,11	10,60	2,18	97,29	98,51	98,70	99,91
South	23 [C°]	0,00	0,00	0,00	10,37	23,55	17,48	10,38
	26 [C°]	0,00	0,00	0,00	44,61	40,73	52,04	42,95
	29 [C°]	8,25	11,28	2,63	96,06	97,46	98,29	98,51
East	23 [C°]	0,00	0,00	0,00	2,95	0,14	8,95	3,85
	26 [C°]	0,00	0,00	0,00	25,24	21,97	34,43	24,50
	29 [C°]	2,13	2,95	0,05	85,29	85,02	93,72	89,44
West	23 [C°]	0,00	0,00	0,00	6,11	0,72	9,36	5,44
	26 [C°]	0,00	0,00	0,00	28,21	23,10	35,93	28,53
	29 [C°]	3,12	5,11	0,27	85,50	84,24	89,14	89,72

In addition, the results of the scenarios of the housing estates located in Mediterranean cities (Algiers, Oran, Bari, and Naples), and which do not have any shading device, are, however, similar between them, and they are good results when compared to those located in the desert climate. Where, the percentages of hours of the comfortable zone (23<To<26) vary between 34.56% in the North, and 22.10% in the West concerning the estates of Algiers; between 17.18% in the south, and 22.38% in the west for those located in Oran; between 34.56% in the south, and 25.48% in the east for the housings located in Bari; and between 34.10% in the north and 20.65% in the east, concerning those of Naples.

It is also observed that the values of the operative temperature simulated in the different scenarios of the apartments located in the Mediterranean cities signal slight variations of the annual percentages of hours in the different zones of discomfort due to the heat ($26<To<29$) for the different orientations of these apartments. Where, the percentage of uncomfortable hours due to the heat of the housings located in Algiers, Oran, Bari, and Naples, reaches an average of 53.19% (± 4.78%) for those oriented to the North, an average of 52.50% (± 4.47%) for the apartments oriented to the south, an average of 61.83% (±2.63%) for the east, and an average of 58.21% (±3.79%) for the housings oriented to the west.

3.2. Comparison of the thermal environment of different housing scenarios with shading devices

To verify the effectiveness of the shading devices in the different housing scenarios in different climates several dimensions of the shading devices were considered 12 cm, 25 cm, and 50 cm. Table 3 shows the cumulative percentages of operative temperature hours of different housing scenarios equipped with shading devices. While Figure 4 shows the hour's percentages of three comfort zones according to the operative temperature of different scenarios of these apartments.

Table 3. Cumulative percentages of operative temperature hours of different housing scenarios equipped with shading devices.

Shading devices	Orientation	Operative temperature	Cumulative percentage [%]						
			Biskra	Ghardaïa	Ouargla	Alger	Oran	Bari	Naples
12 cm	North	23 [C°]	0,00	0,00	0,00	32,92	28,53	36,96	31,98
		26 [C°]	4,21	7,02	0,77	89,71	91,58	92,62	92,67
		29 [C°]	27,26	31,07	14,45	100	100	100	100
	South	23 [C°]	0,00	0,00	0,00	32,74	27,7	35,36	30,66
		26 [C°]	4,12	7,15	0,91	87,18	89,56	91,25	90,27
		29 [C°]	26,76	30,84	14,59	100	100	100	100
	East	23 [C°]	0,00	0,00	0,00	25,13	20,97	27,67	23,72
		26 [C°]	2,13	3,90	0,14	80,93	81,61	86,04	81,92
		29 [C°]	21,34	24,50	11,23	100	100	100	100
	West	23 [C°]	0,00	0,00	0,00	24,63	19,75	27,72	23,96
		26 [C°]	2,54	4,21	0,14	78,53	79,76	84,34	83,06
		29 [C°]	21,61	24,51	11,28	100	100	100	100
25 cm	North	23 [C°]	0,00	0,00	0,00	38,44	34,65	41,93	37,78
		26 [C°]	6,47	9,24	1,99	94,15	95,84	95,73	96,20
		29 [C°]	32,33	34,51	17,52	100	100	100	100
	South	23 [C°]	0,00	0,00	0,00	37,50	33,65	40,12	35,97
		26 [C°]	6,11	9,14	2,04	91,27	94,02	94,61	94,39
		29 [C°]	31,66	34,28	17,12	100	100	100	100
	East	23 [C°]	0,00	0,00	0,00	29,94	25,01	32,38	28,22
		26 [C°]	3,67	5,62	0,27	84,29	86,19	90,21	87,55
		29 [C°]	24,91	27,77	12,95	100	100	100	100
	West	23 [C°]	0,00	0,00	0,00	30,66	24,96	33,20	30,38
		26 [C°]	4,21	6,03	0,27	84,14	85,28	89,54	88,82
		29 [C°]	26,31	27,95	13,18	100	100	100	100
50 cm	North	23 [C°]	0,00	0,00	0,00	51,37	35,55	43,16	39,36
		26 [C°]	6,98	9,51	2,31	96,38	96,29	96,74	96,78
		29 [C°]	33,75	35,60	18,53	100	100	100	100
	South	23 [C°]	0,00	0,00	0,00	50,68	34,65	41,66	37,91
		26 [C°]	6,75	9,51	2,31	95,83	95,30	95,73	95,78
		29 [C°]	33,02	35,20	18,21	100	100	100	100
	East	23 [C°]	0,00	0,00	0,00	43,58	29,94	35,24	31,20
		26 [C°]	4,75	6,52	0,63	91,47	89,85	92,26	91,31
		29 [C°]	26,99	29,71	14,21	100	100	100	100
	West	23 [C°]	0,00	0,00	0,00	44,88	27,71	36,14	32,65
		26 [C°]	5,03	6,84	0,63	91,81	88,68	92,03	91,67
		29 [C°]	27,90	30,58	14,21	100	100	100	100

Regarding the comfortable zone (23<To<26), it is observed that the apartments equipped with the different shading devices (12 cm, 25 cm, and 50 cm), and which are located in the Mediterranean cities (Algiers, Oran, Bari, and Naples) have higher average annual hours compared to the discomfort zones due to the heat (26<To<29) and (29<To). Where, the comfortable zone of the north-facing apartments in the cities of Algiers, Oran, Bari and Naples reaches, respectively, maximum numbers of hours of 56.79%, 63.05%, 55.66%, and 60.69%, for the 12 cm shading device scenario, Figure 4 (a). Also, for those oriented to the south, where they reached, respectively, the 54.44%, 61.86%, 55.89%, and 59.61% of comfortable hours for the scenario of 12 cm, Figure 4 (b). The apartments facing east in the cities of Algiers and Bari reached, respectively, 55.80% and 58.37% of annual comfortable hours for the scenario of a 12 cm shading device, while the percentages of 61.18% and 60.11% were obtained for the apartments in the city of Naples and that have a 50 cm shading device, Figure 4 (c). The percentages of comfortable hours that have been observed by the simulation of the apartments oriented to the west in the cities of Algiers, Oran, Bari, and Naples are similar, where they get 53.90%, 56.62%, and 59.10% for the scenario of 12 cm, while they reach 60.97% for the scenario of 50 cm, Figure 4 (d). However, it is observed by Figure 4 and Table 3 that the different scenarios of the housing located in the desert cities have the lowest average annual comfortable hours, compared with the hours of the discomfort zones due to hot (26<To<29) and (29<To), where they varied between 0.14% and 7.15% with a very low standard deviation of 2.43%, for the different orientations and the different scenarios of shading devices.

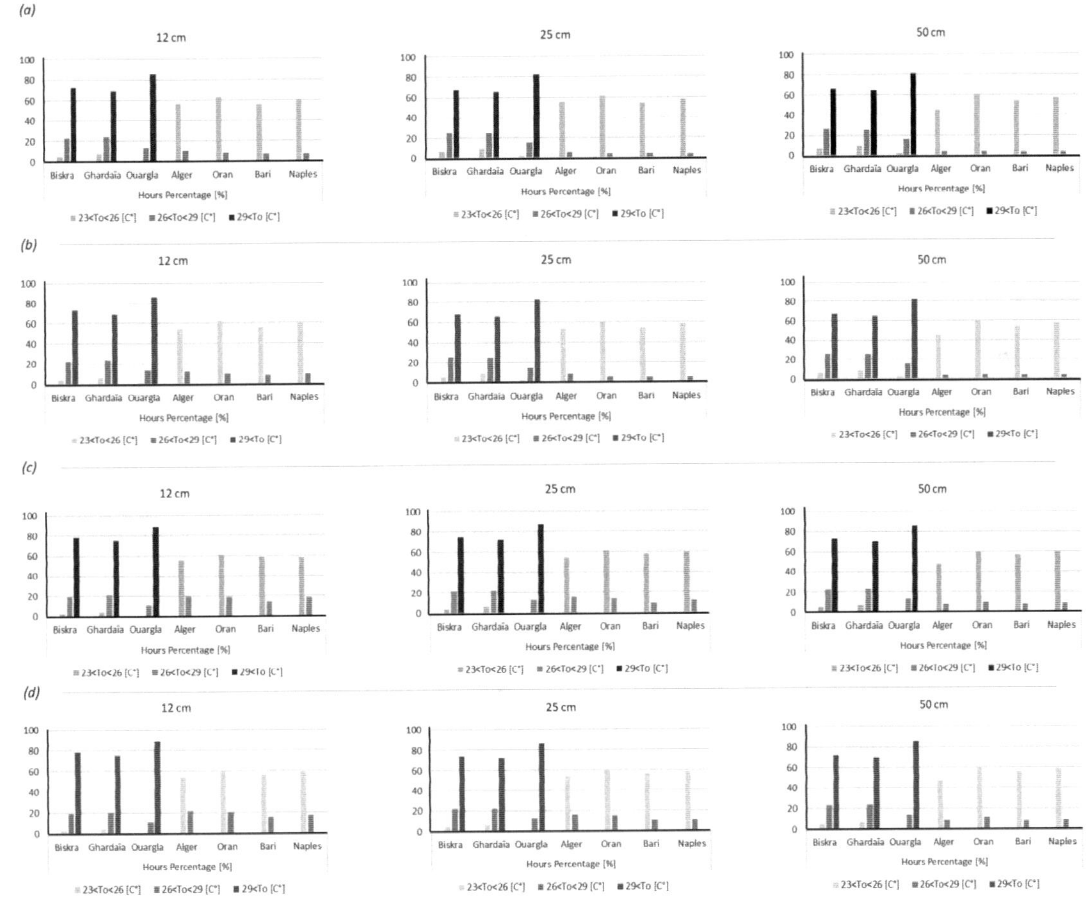

Figure 4. Percentage of operative temperature hours of different housing scenarios equipped with shading devices: (a) North, (b) South, (c) East, (d) West..

IOP Conf. Series: Earth and Environmental Science **992** (2022) 012004 doi:10.1088/1755-1315/992/1/012004

Concerning the very uncomfortable zone due to the heat (29<To), it is observed that the scenarios of the apartments located in the Mediterranean area have no hours in this discomfort zone all the year 0.00%, this means the effectiveness of the shading devices for the apartments located in this region. On the other hand, it is worth noting that the apartments equipped with different shading devices and located in desert cities have the highest average hours in this discomfort zone (29<To), compared with other zones (23<To<26) and (26<To<29), despite the fact that the shading devices cause a slight decrease in the number of hours of discomfort due to the heat, while increasing the scope of these devices from 12 cm to 50 cm. Where, the apartments oriented to the north of the city of Biskra, Ghardaïa and Ouargla reached, respectively, the maximum percentages of hours of the very uncomfortable zone with 72.74%, 68.93% and 85.55% for the scenario of 12 cm, and 65.67%, 65.49% and 82.48% for the scenario of 25 cm, and 66.25%, 64.40% and 81.47% for the scenario of 50 cm, Figure A2 (a). Also, similar results were observed for apartments facing the other orientations (South, East, West). For the South orientation, the percentages of hours of this area uncomfortable housing located in Biskra, Ghardaïa and Ouargla were, respectively, 73.24%, 69.16%, and 85.41% for the scenario of 12 cm, and they were 66.98%, 64.80% and 81.79% for the scenario of 50 cm, Figure A2 (b). While for the east orientation, the percentages of hours of the discomfort zone were, respectively, 78.66%, 75.50%, and 88.77% for the scenario of 12 cm, and they were 73.01%, 70.29%, and 87.79% for the scenario of 50 cm, for the housing located in Biskra, Ghardaïa and Ouargla, Figure A2 (c). While for the western orientation, the hour's percentages of 78.39%, 75.49% and 88.72% were observed for the scenario of 12 cm, and 72.10%, 69.42% and 85.79% for the scenario of shading devices of 50 cm of the apartments located in Biskra, Ghardaïa and Ouargla, respectively, Figure A2 (d).

4. Conclusion and Discussions

This study consists in examining the thermal performance of the envelope components used in the construction of promotional apartments in Algeria and verifying if they are adapted to the desert climate or the Mediterranean climate. A comparative study was made in this research between the results obtained by the simulation of the operative temperatures of an apartment in Biskra, which was looked at as scenarios for the other desert cities (Ghardaïa, and Ouargla) and Mediterranean cities (Algiers, Oran, Bari, and Naples). Where the models were considered as without shading devices (0 cm), and with shading devices of different sizes (12 cm, 25 cm, and 50 cm).

Findings from this research suggest that the indoor thermal environment of apartments that have no shading devices, located in desert cities (Biskra, Ghardaïa, and Ouargla), is defined as thermally aggressive (To<29), where, in which it is not possible to expect the conditions of comfort and health, as well as the occupants, maybe in danger [27] as a result of high indoor temperatures. This explains why the double-wall hollow brick construction system is not suitable for hot and dry climates. In addition, the different scenarios of apartments located in Mediterranean cities (Algiers, Oran, Bari, Naples) report slight variations in the percentages of annual hours in the different zones of discomfort due to heat (26<To<29).

Moreover, the findings show, on the one hand, that apartments equipped with different shading devices (12 cm, 25 cm, and 50 cm) and located in Mediterranean cities have the highest comfort zone (23<To<26), compared to the zones of discomfort due to heat (26<To<29) and (29<To), and on the other hand, that the inhabitants of the Mediterranean apartments do not spend time in a very uncomfortable situation (29<To), because the annual percentage of hours in this zone is 0.00%. From this, we can deduce that the shading devices are effective for housing located in this region. In contrast, the findings show that compared to the number of hours (26 <To <29) and (29 <To) in uncomfortable areas due to heat, the annual averages of comfort hours in different scenarios of housing located in desert cities are the lowest, and although shading devices cause a slight decrease in the number of hours of discomfort due to heat while increasing the reach of these devices from 12 cm to 50 cm, the average hours of the discomfort zone (29<To) in the desert apartments scenarios are the highest. This explains that except for the use of a cooling mechanical system during summer, there is no other solution that can ensure thermal comfort in the desert region [28].

This study is limited in the evaluation of the effect of a single type of composite shading devices (horizontal and vertical) of different dimensions on a living room of an apartment, regarding the different orientations, however, it should be noted that to have a comfortable indoor thermal environment, it is imperative to use the right type of shading devices in the correct period and disposition [29]. Based on the findings of this study, it is believed that future research on data evaluation of parametric simulations of kinetic shading devices is needed to search for the most optimal designs of shading devices for each climate.

5. References

[1]. Del Ferraro, S., Iavicoli S., Russo S., & Molinaro V. (2015), 'A field study on thermal comfort in an Italian hospital considering differences in gender and age', Applied Ergonomics 50: 177-184. https://doi.org/10.1016/j.apergo.2015.03.014

[2]. Steskens, P. W. M. H., & Loomans, M. G. L. C. (2010), 'Performance indicators for health, comfort and safety of the indoor environment', In 10th REHVA World Congress (CLIMA 2010), Antalya, Turkey (p. 9).

[3]. Schellen L., Loomans M., Kingma B., De Wit M., Frijns A., & van Marken Lichtenbelt W. (2013), 'The use of a thermophysiological model in the built environment to predict thermal sensation: coupling with the indoor environment and thermal sensation', Building and Environment 59:10-22. https://doi.org/10.1016/j.apergo.2015.03.014

[4]. Lairgi L., Lagtayi R., Daya A., Elotmani R., & Touzani M. (2021), 'The Impact of Arid Climate on the Indoor Thermal Comfort in the South-East of Morocco', International Journal of Photoenergy 2021. https://doi.org/10.1155/2021/5554629

[5]. Hensen J. L. (1990), 'Literature review on thermal comfort in transient conditions', Building and environment 25/4: 309-316. https://doi.org/10.1016/0360-1323(90)90004-B

[6]. Nicol J. F., & Humphreys M. A. (2002), "Adaptive thermal comfort and sustainable thermal standards for buildings," Energy and buildings 34/6: 563-572. https://doi.org/10.1016/S0378-7788(02)00006-3

[7]. Azizpour F., Moghimi S., Salleh E., Mat S., Lim C., & Sopian K. (2013), 'Thermal comfort assessment of large-scale hospitals in tropical climates: A case study of University Kebangsaan Malaysia Medical Centre (UKMMC)', Energy and Buildings 64: 317-322. https://doi.org/10.1016/j.enbuild.2013.05.033

[8]. Wang Y., Liu Y., Song C., & Liu J. (2015), 'Appropriate indoor operative temperature and bedding micro climate temperature that satisfies the requirements of sleep thermal comfort', Building and Environment 92: 20-29. https://doi.org/10.1016/j.buildenv.2015.04.015

[9]. Bekkouche S., Benouaz T., Yaiche M., Cherier M., Hamdani M., & Chellali F. (2011), 'Introduction to control of solar gain and internal temperatures by thermal insulation, proper orientation and eaves', Energy and buildings 43/9: 2414-2421. https://doi.org/10.1016/j.enbuild.2011.05.018

[10]. Ghedamsi, R., Settou, N., Gouareh, A., Khamouli, A., Saifi, N., Recioui, B., & Dokkar, B. (2016), 'Modeling and forecasting energy consumption for residential buildings in Algeria using bottom-up approach', Energy and Buildings 121: 309-317. https://doi.org/10.1016/j.enbuild.2015.12.030

[11]. Dincer, I. & Zamfirescu, C. (2011), 'Sustainable energy systems and applications', Springer Science & Business Media. https://doi.org/10.1007/978-0-387-95861-3

[12]. Liu J., Kim M. K., & Srebric J. (2021), 'Numerical analysis of cooling potential and indoor thermal comfort with a novel hybrid radiant cooling system in hot and humid climates', Indoor and Built Environment 1420326X211040853. https://doi.org/10.1177/1420326X211040853

[13]. Ghedamsi, R., Settou, N., Gouareh, A., Khamouli, A., Saifi, N., & Recioui, B. (2015), 'Estimating the energy consumption in building sector in Algeria using bottom-up mode', In IREC2015 The Sixth International Renewable Energy Congress (pp. 1-6). IEEE. https://doi.org/10.1109/IREC.2015.7110942

[14]. Sehli, A., Hasni, A., & Tamali, M. (2012), 'The potential of earth-air heat exchangers for low energy cooling of buildings in South Algeria', Energy Procedia 18: 496-506. https://doi.org/10.1016/j.egypro.2012.05.061

[15]. Xiong J. & Tzempelikos A. (2016), 'Model-based shading and lighting controls considering visual comfort and energy use', Solar Energy 134: 416-428. https://doi.org/10.1016/j.solener.2016.04.026

[16]. Noguchi, M. (2016), 'ZEMCH: toward the delivery of zero energy mass custom homes', Springer. https://doi.org/10.1007/978-3-319-31967-4

[17]. Ali, A. A. E. M. M. (2012), 'Using simulation for studying the influence of vertical shading devices on the thermal performance of residential buildings (Case study: New Assiut City)', Ain Shams Engineering Journal 3/2: 163-174. https://doi.org/10.1016/j.asej.2012.02.001

[18]. Berkouk D., Bouzir T. A. K., & Zeroual M. (2020), 'The effect of indoor luminous environment on the comfort of chess players', Journal of science and technics of physical and sports activities 10: 02.

[19]. Berkouk D. (2017), 'Évaluation du confort thermique et lumineux dans le logement collectif: Étude comparative entre le social et le promotionnel, dans la ville de Biskra',Doctoral thesis, Biskra University, Algeria. [Online]. Available: http://thesis.univ-biskra.dz/id/eprint/2972

[20]. Bouzir, T. A. K., Zemmouri, N., & Berkouk, D. (2018), 'Assessment and analysis of noise pollution in Biskra public gardens (Algeria)', In AIP Conference Proceedings (Vol. 1968, No. 1, p. 030069). AIP Publishing LLC. https://doi.org/10.1063/1.5039256

[21]. Ali, A. A. E. M. M. (2013), 'Using simulation for studying the influence of horizontal shading device protrusion on the thermal performance of spaces in residential buildings', Alexandria Engineering Journal 52/4: 787-796. https://doi.org/10.1016/j.aej.2013.09.008

[22]. Berkouk D. & Mazouz S. (2017), 'Numerical study of the horizontal shading devices effect on the thermal performance of promotional apartments: Case of El-Alia-Biskra', Journal of Applied Engineering Science & Technology 3/1: 21-27.

[23]. Berkouk, D., Bouzir, T. A. K., & Mazouz, S. (2018), 'Numerical study of the vertical shading devices effect on the thermal performance of promotional apartments in hot dry climate of Algeria', In AIP Conference Proceedings (Vol. 1968, No. 1, p. 030040). AIP Publishing LLC. https://doi.org/10.1063/1.5039227

[24]. Atef, A., Noureddine, Z., & Soufiane, F. (2015), 'SPUCAL_mrt as a new model for estimating the mean radiant temperature in arid lands', Energy Procedia 74: 273-280. https://doi.org/10.1016/j.egypro.2015.07.603

[25]. Fabbri, K. (2015), 'Ergonomics of the Thermal Environment. Human Body and Clothes', in Indoor Thermal Comfort Perception, 25-74. Springer. https://doi.org/10.1007/978-3-319-18651-1_3

[26]. Kadri N. & Mokhtari A. (2011), 'Contribution à l'étude de réhabilitation thermique de l'enveloppe du bâtiment', Journal of Renewable Energies 14/2 : 301-311.

[27]. Rizzo G., Beccali M., & Nucara A. (2004), 'Thermal comfort', Encyclopedia of Energy 6: 55-64.

[28]. Khoukhi M. & Fezzioui N. (2012), 'Thermal comfort design of traditional houses in hot dry region of Algeria', International Journal of Energy and Environmental Engineering 3/1: 1-9. https://doi.org/10.1186/2251-6832-3-5

[29]. Kirimtat A., Koyunbaba, B. K. Chatzikonstantinou I., and Sariyildiz S. (2016), 'Review of simulation modeling for shading devices in buildings', Renewable and Sustainable Energy Reviews 53: 23-49. https://doi.org/10.1016/j.rser.2015.08.020

IOP Conf. Series: Earth and Environmental Science **992** (2022) 012005 doi:10.1088/1755-1315/992/1/012005

The Influence of Using Responsive Façade as a Tool for Improving the Built Environment: *Case study: Attaba – Opera square*

M F Eltanboly[1] and M M Afify[2]

[1]Architecture Department, Teaching Assistant, Cairo Higher Institute for Engineering, Computer Science and Management, No.2, Banafseg Center2, first Settlement, Cairo, Postcode 11477, Egypt and PhD. Student, Environmental Design and Energy Efficiency Department, Cairo University Faculty of Engineering, Giza

[2]Architecture department, Professor in Environmental Design and Energy Efficiency Department, Cairo University Faculty of Engineering, Giza

Abstract. The path we are currently following to improve the physics of construction due to the destruction of our environment through the negative impact of the built environment around us. Current efforts to decrease carbon and energy are noted by modifying and improving modern "layered" envelopes and improving a misunderstanding of thermal comfort that has become much less effective than expected. Understanding construction physics is an important factor in improving the quality of life indoor and outdoor, relying primarily on theoretical and computational aspects, then applying the practical part to reduce excessive use of carbon and energy and hinder the elaboration of a sustainable and built environment. This paper shows how the problem stems from neglecting to think about the first principles and basic construction physics. Similarly, we show how combining good construction physics with reassessing old methodologies to construction and structure use provides some powerful and effective tools to address the climate emergency. The use of construction physics methods supports the search for new solutions to express the most important factors affecting thermal conduction and minimum surface temperatures. However, buildings have integrated concepts in which advanced systems work together to optimize energy, comfort, and health performance. Convergence between various engineering fields and the overlapping field of construction techniques and services has great potential in the future to achieve the next step in energy saving. As a result, the research highlights new methods that make our buildings interact with the external environment using climate data, responsive envelope design, simulation software, thereby achieving architectural response to surrounding variables, improving building physics to reduce environmental pollution, thermal comfort, and achieving the highest energy efficiency and performance in buildings. This is illustrated by the presentation of a case study of an existing building (Opera Garage Building) – Cairo, it will review solutions to enhance the envelope of the building and its impact on the built environment to come up with some recommendations and solutions which we can apply it on the existing buildings in Egypt.

Keywords: Building Physics- Responsive systems- Building performance- The built environment- Thermal comfort.

1. Introduction

One of the quintessential challenges facing the built environment is its fundamental impact on climate change and the energy sector, as its effects already dominate. Additionally, methods for designing, analyzing, and measuring design solutions of sustainability problems for architects have emerged as pivotal areas of research and development. Building envelope, microclimate, heating, ventilation, and air conditioning (HVAC systems), contribute as main component responsible for energy consumption in buildings. Therefore, sustainable design strategies contain the use of low energy materials for building envelope components, utilizing daylighting to eliminate the need for supplementary artificial lighting during the day, reducing air

IOP Conf. Series: Earth and Environmental Science **992** (2022) 012005 doi:10.1088/1755-1315/992/1/012005

conditioning usage through passive ventilation strategies, and proposing shading devices to solar exposed orientations of buildings [1]. Building energy efficiency is often below average, particularly for its envelope and wall/window shading. For hotter regions, improving room ventilation through better positioning of windows, increasing window size, and using adaptive systems would enhance energy efficiency and achieve thermal comfort for users [2]. Nevertheless, modifying the aspects of building physics and using responsive systems will improve the microclimate because it affects the indoor and the outdoor environment quality.

2. Objectives

The paper aims to explain the importance of applying responsive systems as tool for improving the micro climate and the built environment specially the hydroponic system as a vertical farming and green wall. In addition, developing on evaluation method to examine how the thermal insulation and energy efficiency aspects of building physics are incorporated throughout the building delivery process. Study the influence of Responsive systems and potentials of computer applications on architecture to discuss its role and function as a resilient approach. Moreover, the main goal of this paper is achieving the environment indoor comfort in terms of efficient daylight distribution, natural ventilation and thermal comfort- minimum glare and minimum direct sunlight- through responsive skin design configuration on high-rise office building, using simulation and evaluation design tools.

3. Methodology

It includes a comprehensive review of the building legislation in force at the time of planning of case study. The research tries to prove that responsiveness in architecture can decrease the consumption of energy in buildings and it can improve the microclimate in the built environment. Presentation of global models in some countries that have applied new technologies and systems to generate the external envelope to provide the consumption of energy in buildings. The consequences of the research will be achieved via the following research methodology and case study:

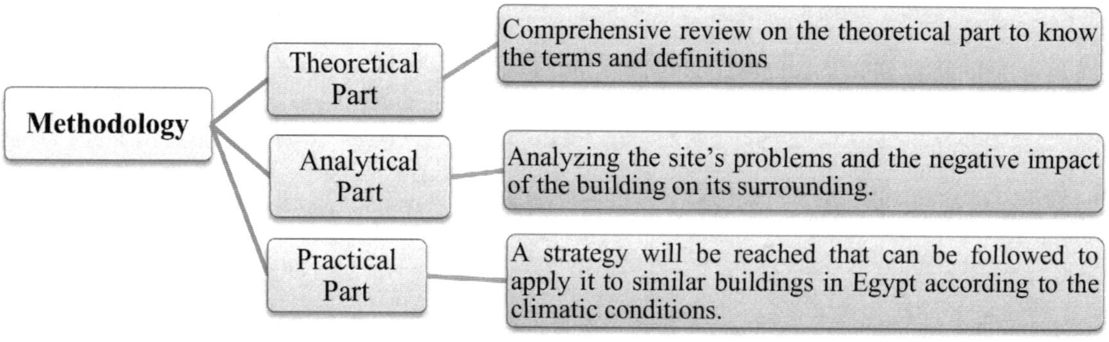

Figure (1): Methodology. **Source:** by author

4. Problem

The problem of research is that there is a lack of awareness about the significance of using natural resources, which has caused energy consumption at large rates, and this constant and speed raise in energy consumption has led to increasing environmental pollution and CO2 emissions, causing global warming and change in climatic phenomena and temperature changes globally. Therefore, the research spots light on the modern design and construction methods that make our cities interact with the external environment by using modern technology and smart materials, thus realizing the architectural response to the surrounding changes [3]. In addition, buildings use an excessive amount of energy to achieve thermal comfort and generate electricity so, thus great quantities of energy are being consumed and the rates of energy consumption are increasing rapidly which causes mainly environmental pollution causing increased CO2 footprint then leading to global warming or climate-changing phenomena. Eco-system becomes imbalance and an increase in environmental pollution and overall global international temperature [3].

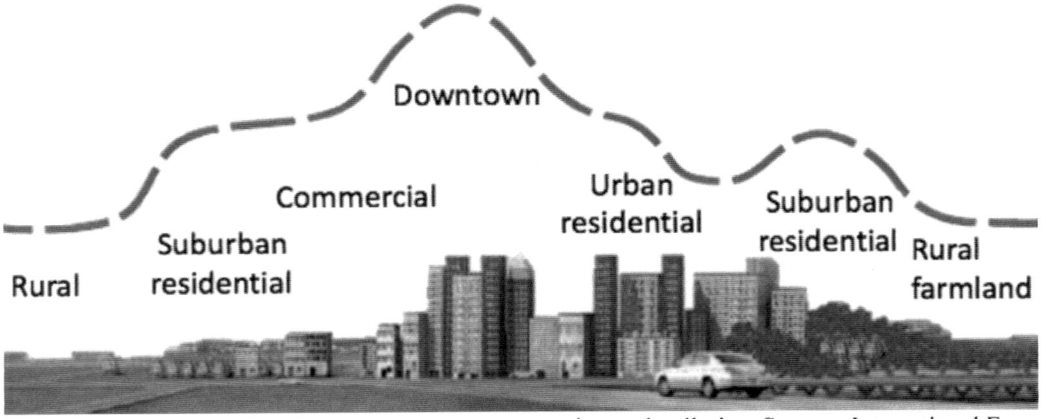

Figure (2): Urban Heat Island increases energy consumption and pollution. **Source:** International Energy Agency

- Buildings are responsible of 45% of the energy consumption in Egypt. (Egypt state of Environment 2006-2007)

Accordingly, the problem is stated as follows; the traditional tools of dealing with complexity which controlling the environmental aspects. As a result, it is essential to consider new technologies and methods in the design process and urban modelling by using responsive systems, computer vision, and simulation applications. This paper shows the challenges in existing building which located in Cairo, this place has many Urbanism and environmental problems such as street vendors, overlapping uses, pollution, and hustle.

5. Questions

This paper clarifies on a lot of questions which in turn can optimize the efficiency of buildings without causing more damage to the environment, like:
- How can the high performance of buildings be achieved through using a responsive systems and modern technology to improve energy efficiency and reach to thermal comfort?
- Is it applicable to apply this technology in Egypt to retrofit the existing building?
- How can the distribution of energy in buildings be controlled by Aerial analysis (Thermal survey) which turns the technology into a powerful, efficient tool?

6. Tools

Simulation programs(Rhino / Envi-met / climate consultant)

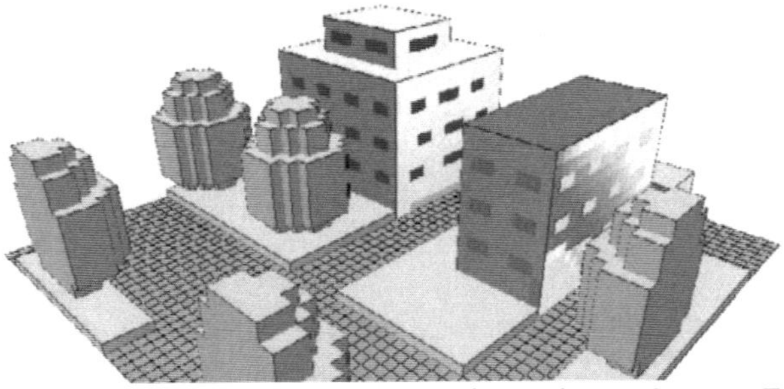

Figure (3): Building Physics simulation analysis via Envi-met. **Source:** Envi-met website

Aerial Analysis (Thermal Survey), Aerial Intelligence for Retrofit Building Energy Modeling (Air-BEM).

Figure (4): Aerial Analysis thermal Survey, **Source**: https://www.svvdroneservices.com/thermal-imaging

7. Literature Review

Systematic literature review:

Building facades have a significant impact on energy use for ventilation in buildings that face warm-humid climates in South East Asian countries [4].The research outlines a three-step process for the design of sustainable facades, as the authors highlight the first step being the analysis of energy consumption, the second step being the reduction of energy consumption by using energy in most efficient ways, and the last step being the fulfilment of remaining energy requirements by renewable energy sources. The research results indicate that following the energy triangle approach has the potential to contribute to sustainable facades in Asia to systematically support building design, which consumes less energy and exploits the potential of natural ventilation [4].

Building envelopes play an important role in global warming and our need to reduce carbon emissions [5]. Their research tried to discover existing visionary building envelope concepts and their relation to maintaining a favourable interior environment. Luther and Altomonte (2007) discuss the concept of biomimicry to combine the most up-to-date building materials and construction technologies for fulfilling the requirements of advanced facades. The authors present building envelopes analogous to nature, physical processes in building envelopes and identify façade as a climate conditioner for increasing thermal comfort. Responsive facades can be designed as vegetation scaffoldings and climate modifiers if integrated with vertical green shading, light shelves, shading devices, and timber screens [5]. The novel contribution of this study included that flexible and environmentally responsible building 'skins' should not be designed as separate components protecting the building interior from external factors, but rather utilize them for integrating advanced materials and systems for optimizing the internal environment according to external conditions also possibly trying to generate energy.

The use of double skin facades has increased significantly over the last 10 to 15 years, primarily due to the benefits attributed to them regarding increased energy efficiency and improved daylighting [6]. There is a debate about whether these benefits would be more effective provided by a well-designed, traditional, single-skin façade system. The research discusses the various types of double skin facade systems and explores their features and function.

The Department of Energetics at Politecnico di Torino conducted various experiments on a climate façade with a mechanically ventilated air gap [7]. The measurements were performed using TWINS (Testing Window Innovative Systems) test facility, which consisted of two outdoor cells, one used for reference purposes, and the other which adopts different active façade configurations. The energy efficiency of the facade and the thermal comfort implications have been evaluated considering the ability to pre-heat the ventilation air in the winter season, and the ability to remove part of the solar load during the summer season; the normalized daily energy passing through the facade and the normalized surface temperature of the inner glass were analyzed [7].

8. Material and Method

Responsive architecture adapts to the surrounding environment causing an alteration in the envelope configuration depends on adaptation. The primary target of a building is to safeguard the inhabitancies from the outer environment and to achieve a comfortable indoor environment. This operation can be reached by the implementation of passive like thermal mass, solar insulation, shading, and natural ventilation. Also, the building was fundamentally designed for the environmental alterations that surround it [8]. Currently, there is a lack of concern regarding the environmental surroundings for the designs of the modern-day buildings; consequently, the building envelope should be designed to be a great potential to save energy and also to react with such alterations. The potentialities of responsive materials and responsive buildings envelopes contribute to the energy decrease.

The responsive architecture enables the architect to design structures that temporary pavilion designed with a stable structure, that can able to adjust to the environmental requests and the changes and needs of the user through the help of suitable technologies [9].

-Responsive architecture: can adapt to corrupting climatic conditions and variables, and that response can be dynamic or static.

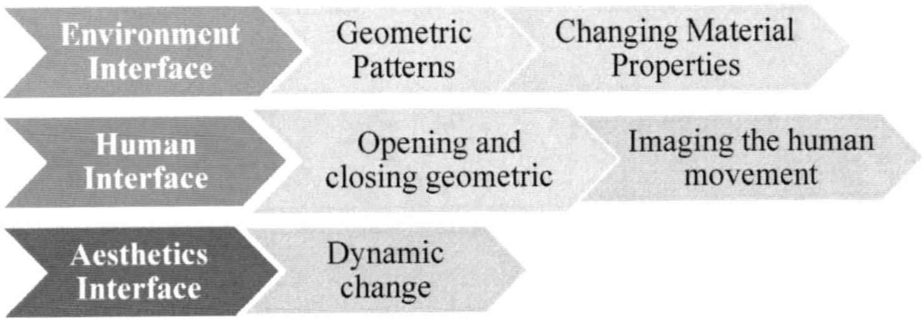

Figure (5): classification of Responsive Architecture. **Source:** by author

Building skins are not considered anymore as an elevation separates inside from outside but as an interactive, responsive element to environment and user's needs, as shown in figure (5). This paper focused on responsiveness to environment interface [3].

The most important factors causing the increase in carbon dioxide and Caused 80% of this problem is generating electricity and the operating Energy inside the building is affected by:

- Building Envelope
- Material Used
- Climatic factors

Although a considerable amount of architecture projects is planned and executed every year and the knowledge about the planning processes is being applied in the field, the integration of building physics aspects in the building delivery process have not yet been extensively documented.

IOP Conf. Series: Earth and Environmental Science **992** (2022) 012005 doi:10.1088/1755-1315/992/1/012005

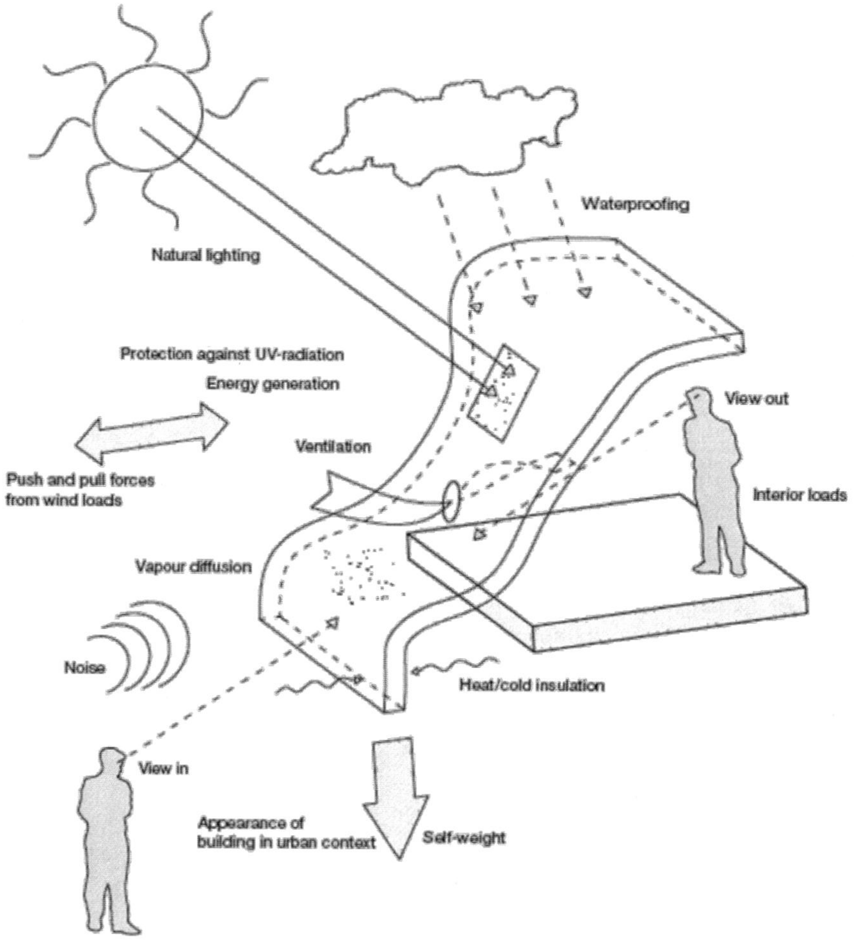

Figure (6): Building Skin Role **Source:** Facades, Knaack, 2007

In contrast, a vast field of studies pertaining to different aspects of building physics can be found. To be able to describe building physics related tasks, it is useful to start by identifying the main building physics aspects in the building delivery process that are relevant for building design, as shown in figure (7) according to design building Wiki (2019) [10]. As a result, understanding building physics is an important factor in improving the quality of life indoor and outdoor, relying primarily on theoretical and computational aspects, then applying the practical part to reduce excessive use of carbon and energy and hinder the development of a sustainable and built environment [7].

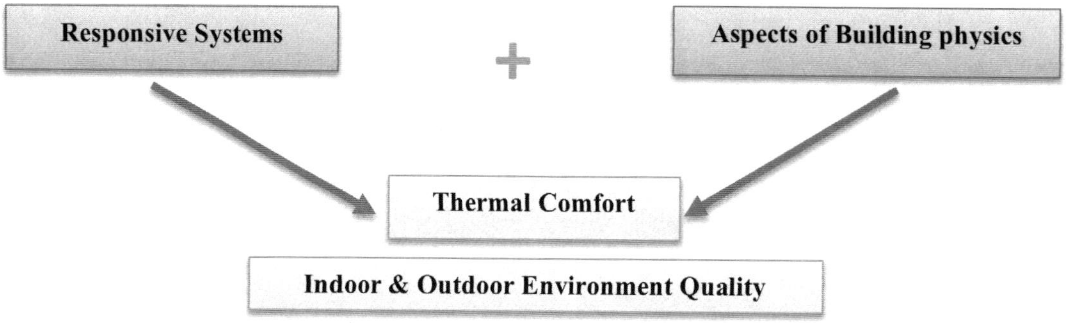

Figure (7): Merging of responsive systems and aspects of building physics. **Source:** by author

The Possibility of applying responsive system in Egypt in Existing buildings: *Case Study – Opera square, application of Hydroponic system (vertical farming):*

The main objective of this project to develop the area of opera square and between the building of the opera garage and Al Attaba and the re-use of the two buildings will positively return to the economic revitalization of the area because it will provide organized places for the street vendors, in addition to that, will revive garages and make them organized commercial spaces in the center of the capital, Cairo. The region lacks entertainment and picnic places for the citizens, which we seek to get rid of by providing recreational places for the people of the region (Environmental aspects). Despite the privileged location of the opera square, as it is located in the heart of Cairo near many commercial centers and markets, in addition to the fact that it faces many challenges that represent obstacles in the way of developing the field and the most important of these challenges:

1- Overlapping Land-use
2- Street vendors
3- Pollution and noise

Figure (8): Attaba Garage, Opera Square

This project aims to create a lung that will serve users and inhabitants in this place to create an urban area where it can be moved and engage to social activities without affecting traffic in the area, the opera square area is characterized by a very crowded and densely populated area and does not have any green areas or gathering points that distinguish that urban area, as there is an opera garage, which is a concrete building consisting of 5 floors but is not used well due to the heavy overcrowding and polluting area.

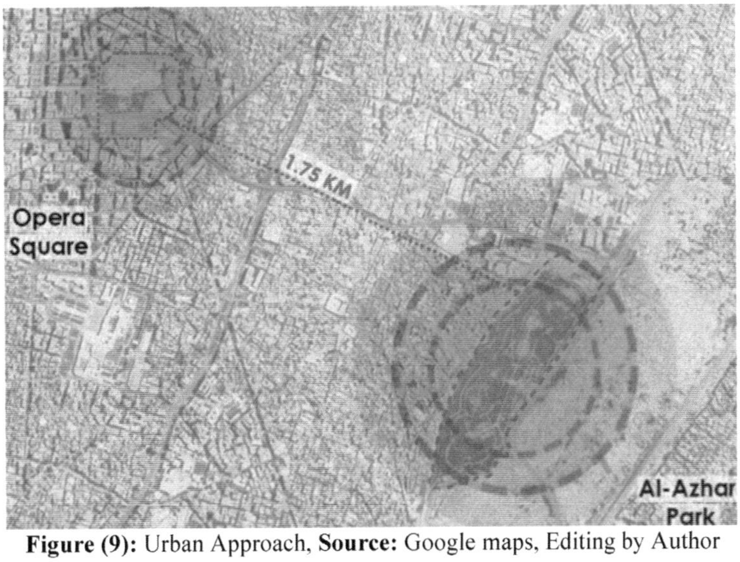

Figure (9): Urban Approach, **Source:** Google maps, Editing by Author

Tools:

- Envi-met Software
- World Weather Data
- Climate consultant for climatic data

This steps illustrate the site analysis data and simulation for the existing situation by using Envi-Met Software to know thermal comfort indoor and outdoor and the effect of building envelop to reach to the outdoor and the indoor environment quality.

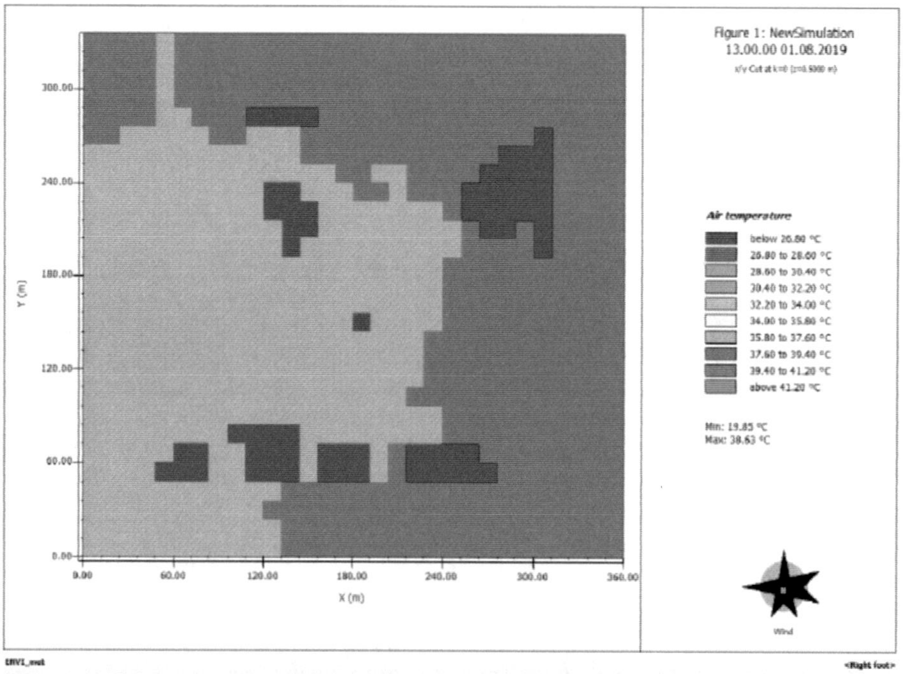

Figure (10): Temperature analysis, **Source:** by Author, using Envi-met software

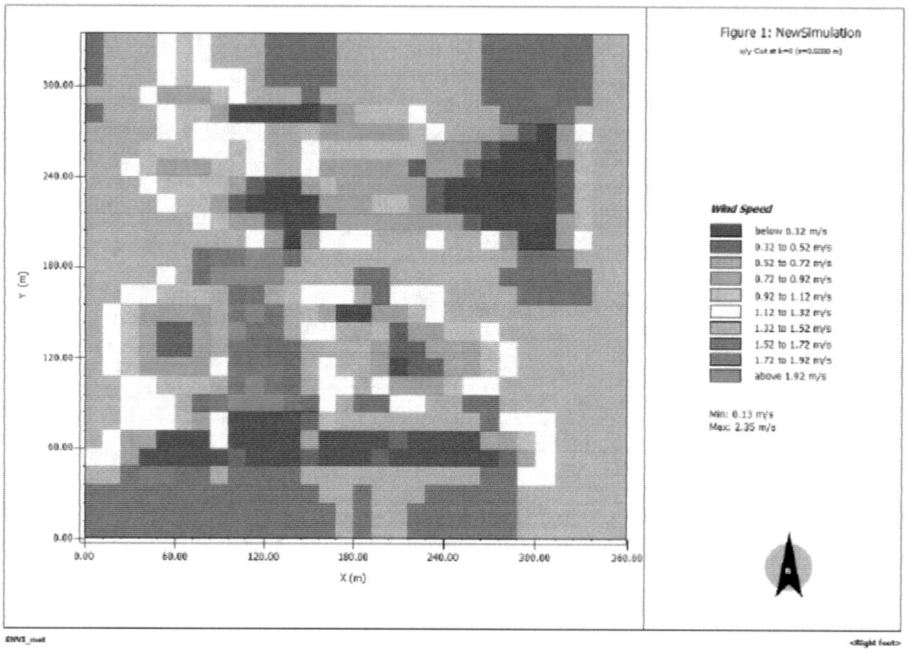

Figure (11): Wind analysis, **Source:** by Author, using Envi-met software

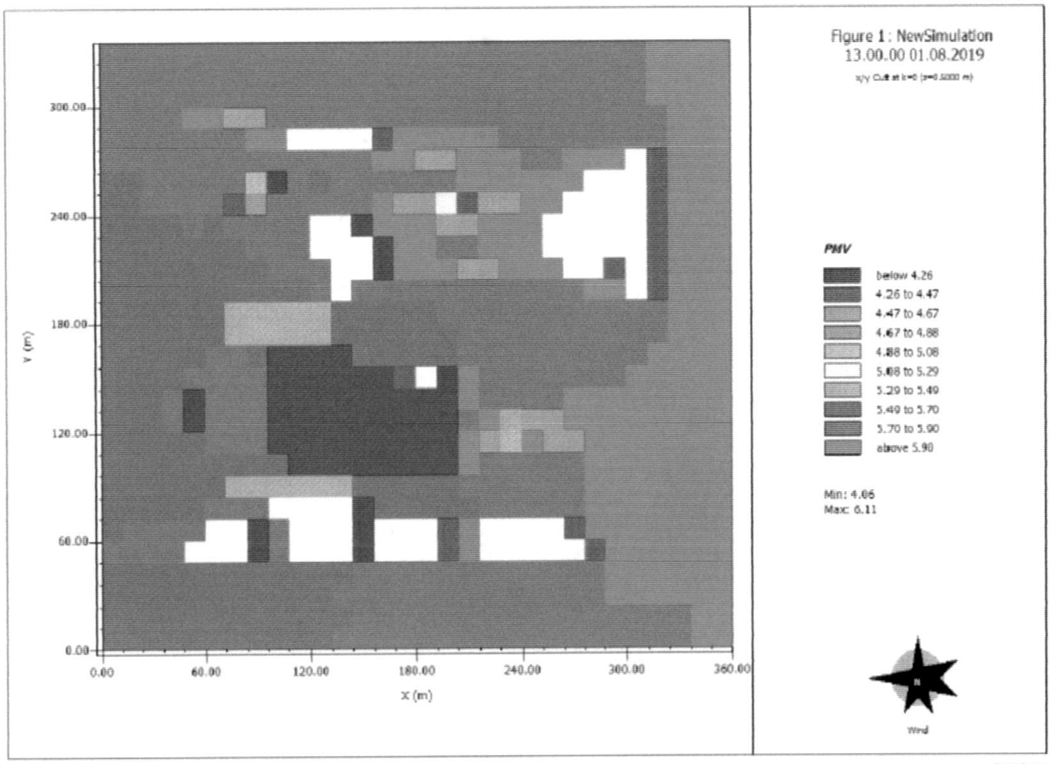

Figure (12): Predicted Mean Vote (PMV) analysis, **Source:** by Author, using Envi-met software

Atmosphere	
Start of simulation (h)	13:00
Wind speed at 10 m above ground level [m/s]	0.32 to 1.12
Air temperature	39.40
Predicted Mean Vote (PMV)	4.06

TABLE 1. Input configuration data applied in the ENVI-met simulations.

This table explain the maximum and minimum degrees and ratios for temperature, wind, and predicted mean vote to be able to reach to thermal comfort. Results show that this area need trees and green spaces to reduce the effect of CO2 emission and the high temperature. Convergence between various engineering fields and the overlapping field of construction techniques and services has great potential in the future to achieve the next step in energy saving. As a result, the research highlights new methods that make our buildings interact with the external environment using climate data, responsive envelope design, simulation software, thereby achieving architectural response to surrounding variables, improving building physics to reduce environmental pollution, thermal comfort, and achieving the highest energy efficiency and performance in buildings. This is illustrated by the presentation of a case study of an existing building (Opera Garage Building) – Cairo, it will review solutions to enhance the envelope of the building and its impact on the built environment to come up with some recommendations and solutions which we can apply it on the existing buildings in Egypt.

IOP Conf. Series: Earth and Environmental Science **992** (2022) 012005 doi:10.1088/1755-1315/992/1/012005

The main findings of this study are:
- The responsive façade reduces energy consumption and carbon dioxide emissions, depending on the climatic condition of the building.
- The allowance of entering the natural light inside the building depends on the material of responsive units and climatic analysis.

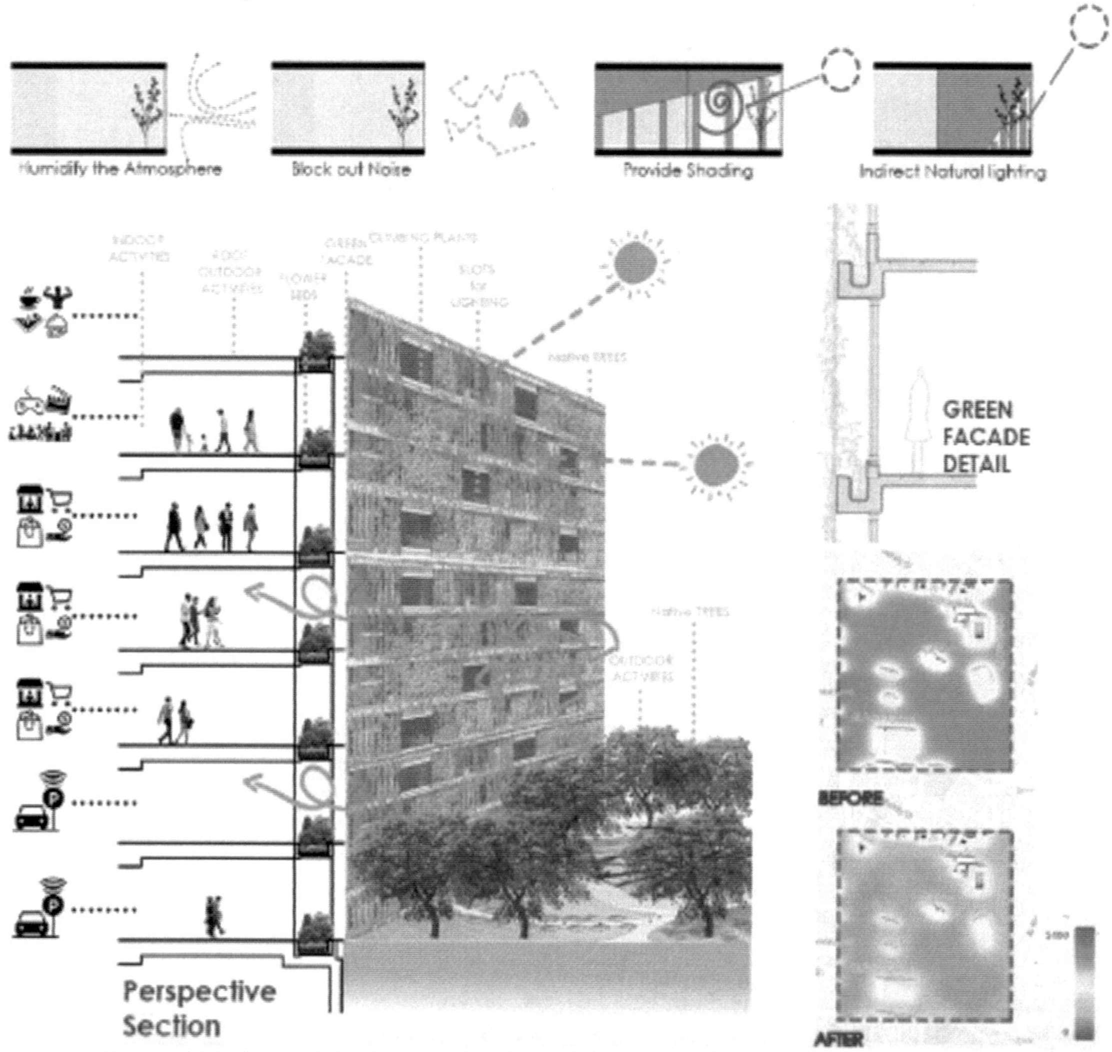

Figure (13): Proposal for the main façade applying vertical farming, **Source:** by Author

Conclusion:

The world is facing a quintessential problem characterized in the lack of energy and pollution related to the ever increase in energy consumption. The buildings contribute to these problems as they are allowed the main consumers of energy and contribute to increasing environmental contaminate. Designing buildings to achieve thermal convenience is associated with studying the mechanisms of heat transfer between the building and the outdoor environment. The main findings of this study are:
- The responsive façade reduces energy consumption and carbon dioxide emissions, depending on the climatic condition of the building

- The responsive systems have a positive effect for improving the microclimate, and it's important to understand the building physics elements to decrease the negative effect of the built environment.
- In general, windows are considered the weak point in the cover of the building in terms of thermal insulation, but with the rapid progress of technology recently, especially in the field of materials engineering and nanotechnology, there is glass with advanced features and highly advanced performance such as glass with vacuum insulation equivalent to isolation insulation of a wall consisting of several layers, and glass It can generate solar power while it is transparent, intelligent glass can be controlled by its transparency and wireless heat acquisition automatically or through a smartphone, and many other technologies [11]

Recommendation and Future Studies

The following are recommendations for the architects and government in order to enhance the application of responsive systems in Egypt specially in existing buildings which consume a large amount of Energy:

- The direction of responsive facade, whether it is vertical or horizontal.
-The mechanism of movement which should not be a complicated movement in order to avoid the system's failure.
- The importance of study site analysis and climatic condition of the area and building to know which a suitable system to use
- Study ways of integrating the building, and maximizing energy efficiency and improving the building's thermal performance.
- Knowledge into a design strategy that assist designer into the design process in making performance based into integrated application in climate-responsive building elements.
- The role of responsive systems as a means of solving the energy consumption problem in buildings.
- The role of computational design to change all design processes and environmental analysis and solve the problems of energy consumption in buildings.

References

[1] Anthony Nkem Ede, Douglas Kendyson, Solomon Olakunle, John Oluwafemi. "Study of Energy Efficient Building Design Techniques: Covenant University Health Centre", Journal of Physics: Conference, Series, 2019.

[2] Daniel Grünkranz, 2015, Towards a Phenomenology of Responsive Architecture: Intelligent Technologies and Their Influence on the Experience of Space Mag. Master of Architecture, PhD Student at the University of Applied Arts, Vienna.

[3] Manar Eltanboly., Ahmed Radwan., Vitta Ibrahim., Analytical Study of Responsive Architecture and Smart Materials and their Impact on Environmental Design, Master thesis, Helwan University, 2019, pg.2,14.

[4] Haase, M., & Amato, A. (2006, September). Sustainable façade design for zero energy buildings in the tropics. In Proceedings of The 23rd Conference on Passive and Low Energy Architecture, Geneva, Switzerland.

[5] Luther, M., & Altomonte, S. (2007, January). Natural and environmentally responsive building envelopes. In 37th International Conference on Environmental Systems (ICES), Chicago, Illinois, July 9-12, 2007. (pp. 1-12). Society of Automotive Engineers.

[6] Pollard, B.(2009). DOUBLE SKIN FAÇADES–MORE IS LESS? Environment Design Guide,1-10

[7] Serra, V., Zanghirella, F., & Perino, M. (2010). Experimental evaluation of a climate façade: Energy efficiency andthermal comfort performance. Energy and buildings, 42(1), 50-62.

[8] Modin, H. (2014). Adaptive building envelopes. Chalmers University of Technology, 18–19.

[9] Sharaidin K., 2014, "Kinetic Facades: Towards design for Environmental Performance", PhD Thesis, School of Architecture and Design, Royal Melbourne Institute of Technology University, Australia.

[10] Frontczak, M., Schiavon, S., Goins, J., Arens, E., Zhang, H., & Wargocki, P. (2012). Quantitative relationships between occupant satisfaction and satisfaction aspects of indoor environmental quality and building design.

[11] Fouad, M. M., Ibrahim, V. A. R., & Radwan, A. H. (2019). The Impact of Responsive systems on energy consumption and thermal performance of buildings. Journal of Al-Azhar University Engineering Sector, 14(52),1037-1049.

IOP Conf. Series: Earth and Environmental Science **992** (2022) 012006 doi:10.1088/1755-1315/992/1/012006

Digital Interactive Installations:
Guidelines towards Social Interaction in City Centers.

Eman H Ayad[1] and Hayam M Omayer[1]

[1]Assistant Professor, Architecture Department, Cairo Higher Institute for Engineering, Computer Science and Management, No.2, Banafseg Center2, first Settlement, Cairo, Postcode 11477, Egypt.

E-mail: emanayad79@gmail.com

Abstract. City center spaces, especially in Egypt, suffer the absence of an interactive visual image and the richness of artistic installations, while its existence totally changes space perception and brings it to life... Although city centers include bunches of architectural elements with multiples vocabularies, yet still most architects and designers miss to include people or space visitors in consideration. Many designs lack the potential for social interaction and participation... This paper explores the role of "Digital Interactive Installations" as a main element in city centers' design, it allows visitors to interact, participate and take part in the formation of a socially amusing experience. Thus, the research aims to identify the characteristics of "Digital Interactive Installations" as an essential tool for city center activation, recommending and guiding designers and decision-makers to achieve social interaction in city centers.

Keywords: Digital Interactive Installations, Social Interaction, Urban Space, Digitalization.

1. Introduction

Interactive installations are artistic configurations designed with high technologies to take part in attracting people, motivating them to interact and participate within its located area [1]. It engages people with the artwork by being part of its operation, and creates a human-computer and a human-human relationship in space. As new technological applications in all various fields of life are continuously accelerating, it became very difficult to activate urban public spaces in cities unless accommodating distinctive installations that mix new various sciences from engineering, art and creativity, etc. These installations allow people to use all their senses to interact; either by moving the installation, or moving through it, controlling its lights & sounds, or promote participation in the space itself. They ensure a dynamic mutual relation with a mutual influence on the user and the object [2]. Public Spaces including interactive installations provide new experiences coping with technology and fulfill people's needs and expectations.

Digital Installations are kind of artworks designed with new media technologies; they include computer graphics, animations, Internet & virtual art, video games, robotics, 3D printing, etc. [1]. They are installations created by digital technology. Moreover, as not all interactive installations are digital, also not all digital installations are interactive. Thus, the target was to combine characteristics of being 'digital' and 'interactive' in installations in order to ensure social attraction and gain high levels of interactions in space. Digital Interactive Installations (DII) provide opportunities for several kinds of interactions between users, the media application, the artwork, the urban space and finally the surrounding physical and climatic environmental features. These Installations best fit to play their role in public city centers or squares, where high levels of social community attraction and interaction are required. A vibrant public city center for today must offer an attractive, comfortable, and secure physical environment, not only through diverse mix of uses, but also by engaging people in space addressing their minds and senses, and where Digital Interactive Installations (DII) must be included. Lately, DII are considered tools for changing public spaces bringing them to life, making them more functional, more attractive and supportive to constantly innovative technologies, thus achieving high quality of urban life.

2. Objectives

The paper aims to explore the distinguished role of 'Digital Interactive Installations' (DII) in attracting people and creating vibrant public spaces, which we urgently need to sustain in city centers, especially throughout pandemic conditions that act as a social interaction repellent.. Therefore, the study focuses on Installations offering high levels of interactivity, which are presented in Interactive and Participatory forms or installations. It specifically addresses 'Digital Interactive Installations' (DII) as a tool to provide an exceptional experience for people in public spaces. It aims to conclude a set of guidelines to be followed for designing a successful DII in city centers, and achieving high levels of social attraction and participation in such spaces.

3. Methodology

This research is an exploratory study that depends on two main sectors. The first is a literature review presenting Interactive Installations; their importance and the different levels of interactivity. It then discusses definitions of Digital Interactive Installations (DII), and focuses on its benefits to determine four main aspects being considered later as main criteria for the study. This sector also presents some recent applications or examples of interactive and participative installations that successfully play a crucial role in the activation of its context, whereas applications are described and analyzed referring to the main four previously mentioned criteria. The second part of the research presents the analytical study that depends mainly on a questionnaire addressing professionals and experts through questions to verify the determined criteria. Finally, from these two main parts, the research concludes a set of guidelines to be considered in designing DII, for achieving a vibrant, attractive city center space.

4. Literature Review

4.1. Interactive Installations

Installations in public spaces have always been subject to change and develop according to the evolution of materials being the main language of artworks [3]. Despite this development, installations stayed static until technology was introduced to public art creating dynamic forms... This technology interference brought 'interactivity' to artistic installations, and with further technology development, computer generated forms depending on smart materials, mobile networks and digital media appeared ensuring and strengthening user interaction... Thus, interactive installations are the dynamic forms of art that depend the spectator's interference aiming to evoke a space-related experience. Spectators than become part of the installation using their senses to operate it.

4.2. Interactivity Levels

Many researches presented concepts of interaction according to different study fields. Considering installations as an artwork could help designers categorize this subject under generative art and technology, where forms could be classified based on the 'degree of interaction' between the form and the viewer as follows: [3]

1- *Static forms…* as statues, with no interaction and no response to context.
2- *Dynamic forms…* are based on certain mechanisms to move or change depending on an environmental change in temperature, light, etc. or a change in time, but the viewer behavior has no impact on the form.
3- *Interactive forms…* they respond to human behavior, where the viewer plays an active role using body language to be captured by sensory systems in the dynamic form, creating a dialogue controlled by human actions and behavior.
4- *Participatory forms…*where interactive forms are transformed to be art platforms offering space for certain activity based on its physical and digitalized systems. The viewer, who eventually became a user contributing to the artwork, interacts and participates producing a dynamic media reflecting his behavior and related to the environment.

IOP Conf. Series: Earth and Environmental Science **992** (2022) 012006 doi:10.1088/1755-1315/992/1/012006

4.3. *Digital Interactive Installations*

- *Definitions*

A special kind of intersection between Art and Technology is presented in 'Digital Interactive Installations', whereas the term 'digital' refers to the use of electronic resources, devices and systems that store, process and produce data in numerical formats[4]. Internet applications, social media, mobile devices, etc. are all products of digital technology. In design and construction, digital data is used to present a user-defined product or result; it provides tools to enhance, support, and efficiently operate the built environment[5]. On the other hand, the term 'interactive' reflects a mutual dynamic response between two or more items, in our subject, between the installation, users and the surrounding context. An 'interactive object' usually shapes an event, it provides space for a special activity to be performed by users, whose behavior or response brings an 'artwork-event' to life[6].

- *Benefits & Criteria Classification*

Digital interactive installations in cities today are having huge impacts on how people engage with public space. A variety of benefits are gained from using these installations in public spaces. Main parties affected are the space visitors, the surrounding activities, the digitalization method, and finally the urban space itself. This part explores and clarifies the benefits to the main mentioned parties being considered as the scale for further study analysis and criteria.

a) *Space visitors...*
   - DII allows visitors to step in, respond to; act & react with the installation in their own way, and some leave their traces behind reflecting various experiences.
   - It facilitates the process of handling technology, providing multi-functional tools, increases interaction possibilities to a great extent.
   - Visitors take action together, in small and large groups, they interact and participate, each being part of an experience. Thus enhancing social connectedness.
   - DII accepts input digital data from visitors to take part in community social networks, thus strengthening active social participation.

b) *The urban space...*
   - As a prime impact of digital technology, DII improves contextual performance and increases productivity, thus making spaces more functional.
   - Urban spaces with DII are more attractive, enjoyable and entertaining, and so more vibrant.
   - DII applies novel technologies to urban spaces changing the whole urban perception to cope with the global rapid technological development.

c) *Surrounding activities...*
   - DII enhances performance of existing activities, attracts brands, thus, improving investments.
   - More visitors are attracted to activities in such spaces, whereas people always attract people.
   - It creates new patterns of social activities, some between people and other within the installation platform or operation.

d) *Digitalization...*

- Bridging the bond between human and digital technology in space give more opportunities for creative ideas considering social interaction, with endless tools for implementations.
- The ability to blend real and virtual experiences with different patterns of activities guarantees efficiency of operation, maintenance and management.

4.4. Applications

The following table presents selected examples for interactive and participative installation, where some operate through digital technology and others are limited to certain mechanical movement or operation… Examples are being analyzed according to location, concept or operating design description and finally the impact of each installation to the four parties (previously determined) as main elements affected by the installation: space visitors, urban space, surrounding activity and digitalization.

Table 1. Selected examples for interactive and participative installation.

	Design description	Impact
"Prismatica" **Location:** Place des Festivals, Montreal 	*Interactive Installation* 50 pivoting prisms made from panels laminated with a dichroic film, transmitting and reflecting every color in the visible spectrum, illuminating visitors at night with colorful lights. At the base of each prism is a projector that creates an infinite interplay of lights and reflections. And, during rotation, a soundtrack of bells plays, too.	- **Space visitors**: excitingly illuminated & interacting to produce different colors, lights & sounds by rotating the prisms. - **Urban space:** succeeded in attracting people offering a space with colored-shifting patterns and reflections [7]. - **Surrounding activities:** attracting more investments to the district [8]. - **Digitalization:** not applied
"BruumRuum" **Location:** Plaza de las Glorias, Barcelona  	*Interactive Installation* A playful light/sound intervention that invites visitors to participate and produce different sound intensities reflecting urban life in this sector. 522 LEDs, embedded in a 3,300sqm ground area, react to the intensity of voices and the ambient sound that the city generates through sensors installed around the plaza... .	- **Space visitors**: all visitors (old-young) play with sounds, highly increasing their sense of belonging and safety [9]. - **Urban space:** turned into a glorious game where city life sounds become a light-colored jazz space. The space offered an atmosphere with a specific emotional 'load'. - **Surrounding activities:** different patterns of activities created with high interaction levels. - **Digitalization:** Limited to sound sensors that transforms sound of people and city to different light patterns.
"Sonic Onyx" **Location:** School courtyard, Trondheim, Norway	*Participative Installation* 3D sculpture about four meters high and seven meters in diameter. The	- **Space visitors**: are attracted by 3D sound to interact with the system through their mobile

 	'legs' are metal and static with a top globe of diameter 1.75m changing its color and intensity. The color depends on the surrounding temperature and the intensity of the light changes according to the speed of the wind. [10] Additionally, 3D sound is played all the time attracting visitors whom interact digitally by their mobile phones…The content sent is processed by the software and played by the sculpture sound system. This includes conversion of images and text into sound and combining it with the background music. The music played is archived and played again later in a random way.	phones. - **Urban space:** the installation enhances the school yard with its aesthetics value. - **Surrounding activities:** students are engaged and involved in learning activities and practicing music - **Digitalization:** Installation depends on a central server, a Bluetooth server, sound system and light system… Digital communication via Bluetooth where the system accepts messages and files from visitors and produces music [11].
"Moon rising from sea" **Location:** Center square, Taicang, Shanghai. 	***Participative Installation*** A public media art installation with a base of 10x10 meters and 8 meters high. The base walls have different reliefs and display screens, and on the top a large sculpture with a shape of a sail and a moon rising from the waves. The sail have reliefs reflecting the city's sea culture, and the moon is used for mounting projectors. Videos, Images and digital applications can be projected on the walls and the inner surface of the sail. Different techniques are used to correct media projected on the curved surface of the sail… Additionally, sound systems are also being installed.	- **Space visitors**: share their photos from social media, creating social connectedness. Citizens working overseas are proud to show others their culture and what they do to the world. - **Urban space:** explores the historical values of the port and its global importance bringing back heritage and cultural impacts. - **Surrounding activities:** Maximum social attraction and connectedness was able to revalue activities to a high extent. - **Digitalization:** a platform for presenting digital media through visitors' participation either by a tangible steering wheel or their own mobile phones.
"Disobedience" **Location:** Biennale, London.  	***Interactive Installation*** Although not permanent in space, but had a great effect during its existence. This installation, designed by Nassia Inglessis (Studio INI team), is 56 feet long, about 10 feet tall, and can expand up to 5 feet. It aimed to challenge sculptural static art forms… It swamps its visitor when getting to walk through it. According to an ancient Greek mythology, Disobedience seeks to transform space visitors from obedient spectator to a disobedient actor. When passing through it physically, the wall skin transforms in response creating an undulating walkway [12].	- **Space visitors**: go through an emotional journey that goes from frustration, curiosity to excitement and discovery. - **Urban space:** a livable space dynamic, emotionally active space, attracting more people. - **Surrounding activities:** creating new patterns of human interactive activities between different visitors and users. - **Digitalization:** 'Augmented Materiality' not purely digital...

Source: Prepared by authors based on references (6, 7, 8, 9, 10, 11, and 12).

5. The Survey Study

This part of the study presents the analytical study that depends on an expert-oriented questionnaire; the questionnaire aims to examine and verify the determined criteria for a successful DII design... Questions were designed based on the specified criteria and research-related studies (e.g. human-computer interaction, digital design, and digital installations art) as shown in the following table.

Table 2. The relationship between questions and the research-related studies & criteria.

	Q1.	Q2.	Q3.	Q4.	Q5.	Q6.	Q7.	Q8.	Q9.
Design concept	✓	✓	✓		✓	✓			✓
Space visitors		✓	✓	✓	✓		✓	✓	✓
Urban space		✓	✓	✓	✓		✓	✓	✓
Surrounding activities			✓	✓					✓
Digitalization		✓	✓		✓	✓	✓		✓

The study employed a (non-probability) sampling technique, specifically (expert sample) [13]. Responses were analyzed by descriptive statistics due to its sensitivity and difficulty, and so the study community needed to be experienced. The questionnaire considered multiple academic architects from various Egyptian universities and some practicing architects. The research used the contingency tables to tabulate the categorical data and the four Likert scale questions. The most important outputs were presented in the form of Excel graphs for visual analysis and better understanding.

6. Results and Discussion

With the aid of a random online questionnaire serving as a primary data collection method, the study was able to conclude the proposed design guidelines. Response rates were calculated and presented in Excel tables and charts. The questionnaire was shared with 100 participants, where the total responses were 43% including practical and scientific expertise from architects and urbanists... Academic architects were 25.6%, Practical architects were 30.2%, and 44.2% were both Academic and Practical architects... According to the results, the highest respondents have more than 15 years of experience with 62.8%, those from 11-15 years were 14%, and those from 5-10 years were 23.3%.
Considering the first two exploring questionnaire questions, results of the statistical analysis are analyzed as follows:

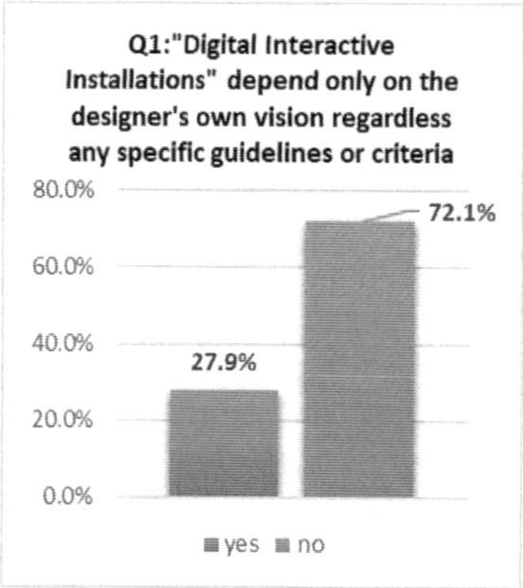

Figure 1. The result of Q1

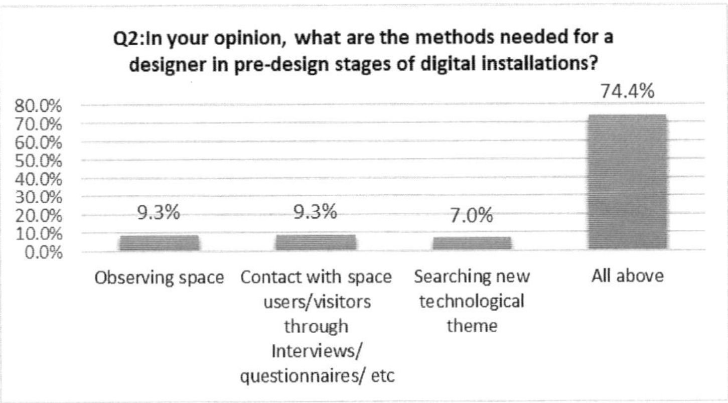

Figure 2. The result of

Questions reflecting DII design concepts and methods

Due to the emergence of new generations of "Interactive Installations" in many countries and their integration with digital applications, designers were pushed to focus on new technological ideas and experiences rather than on integrated design guidelines, as the sample agreed...

Referring to Q1 & Q2, some experts believe that:

- It's a comprehensive process, dependent on stakeholders according to the community culture, needs and space requirements. It also relies on many physical, financial, social, etc. aspects. They need innovation that reflects the smartness of the intelligent process.
- Some experts believe that each city has unique goals, objectives, and visions from other cities and so DII needs to be integrated and associated with these visions, like to have more specific design criteria and pilot studies.
- A designer's vision and standards complement each other... In addition to the designer's vision, the design process should be based on specific criteria for the user and built environment, preferably based on criteria for attractions, the way light and sound are used, and the configurations that help to show the design element in its best image.
- Another dimension mentioned was that the designer had to consider disability, health conditions, age, etc.
- In the view of some, there are no guidelines or criteria for digital Interactive Installations. Therefore, techniques are associated with their own criteria; they have to be taken into account.

Questions reflecting DII design considerations

One of the research objectives was to explore the main design considerations of DII. As per the findings, illustrated in figure 3, most experts believe that attracting people and enhancing their experience, along with the availability of the required technology are the main design considerations in the design process with an average of 27.9%, followed by visitors' interaction & participation with 23.3%, while promoting new ideas and concepts reached 20.9%. Some experts see that there are other design considerations, such as cultural and traditional practices, visitors' needs, people's awareness of innovation concepts, and to be appropriate to the nature of the local reality, in addition to the character and identity of the place.

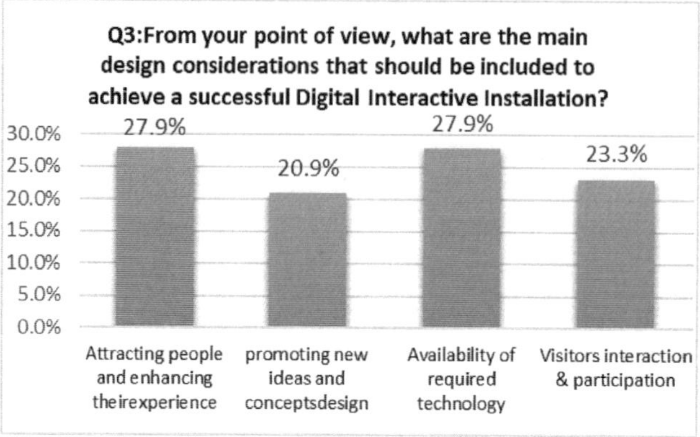

Figure 3. The result of Q3

Questions reflecting DII benefits for determined criteria

Referring to the results of Q4, the most important benefit that a digital interactive installation should offer in a city center space is as follows:

- **To people**, allowing visitors to participate in new social activities is viewed as a critical benefit by 30.2% of the sample, while 25.6% think that offering new opportunities for visitors to participate actively is the main benefit, followed by 23.3% for changing public perception, and lastly, 20.9% for encouraging visitors to take part in community social networks... An obvious example was the Barcelona "BrummRumm" where visitors are main actors for an extremely enjoyable and amusing experience.
- **For the urban space benefits**, the higher percentages were 41.9% for making spaces more enjoyable and attractive, and 32.6% for improving space performance, and 16.3% for applying technology to urban life.
- **To the surrounding activities,** according to the results, this question found out that 41.9% of the experts chose to improve the performance of activities as the most important benefit, against 30.2% for creating new patterns of activities. The lowest percentages were 14% for attracting more visitors to activities and increasing the rate of investment... As mentioned in the above applications, the London Biennale "Disobedience" Installation aimed to immerse visitors into a discovery journey that creates new patterns of interaction.

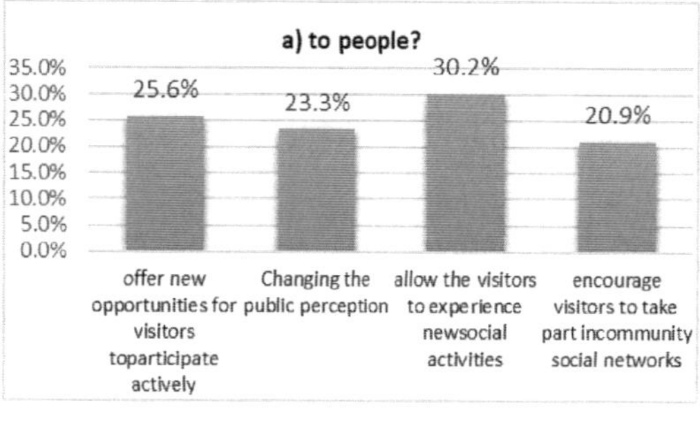

Figure 4. The result of Q4, part A

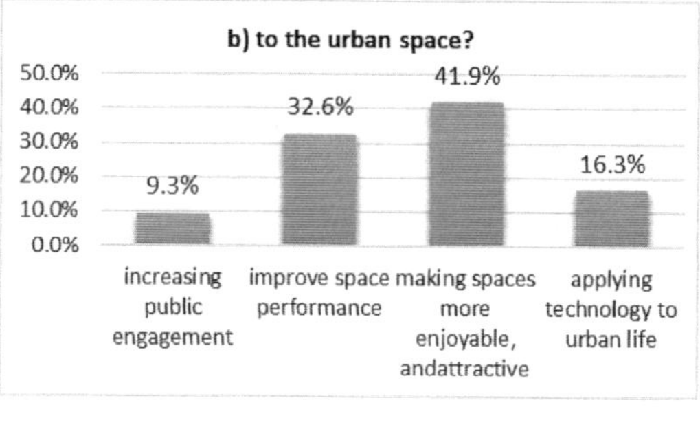

Figure 5. The result of Q4, part B

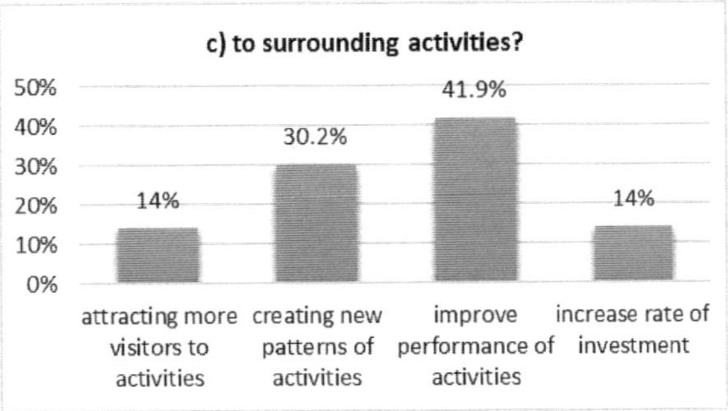

Figure 6. The result of Q4, part C

Questions reflecting DII challenges

The fifth question (Q5) aims to explore the challenges that face the designer in achieving a successful DII design in city centers. A percentage of 27.9% think managing and maintaining the installation is a big challenge for the designer, which was the case in the "Sonic Onyx" installation (previously mentioned) where maintenance problems such as hardware and configuration related problems appeared [12], while 25.6% believe that obtaining people's attention and interaction is the big challenge, followed by 23.3% for urban regulations and constraints by authorities. By asking experts if there are other challenges that haven't been mentioned, some of them added that the cost is an important issue that can negatively affect the design. Furthermore, some mentioned that society's culture and awareness towards digital installation is a great challenge in promoting the idea and getting people to interact, while other ones think that providing training and expert staff in digitalization is a big challenge that can face a DII.

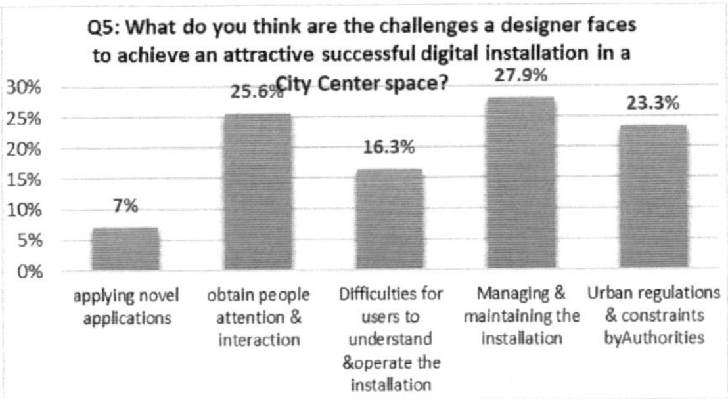

Figure 7. The result of Q5

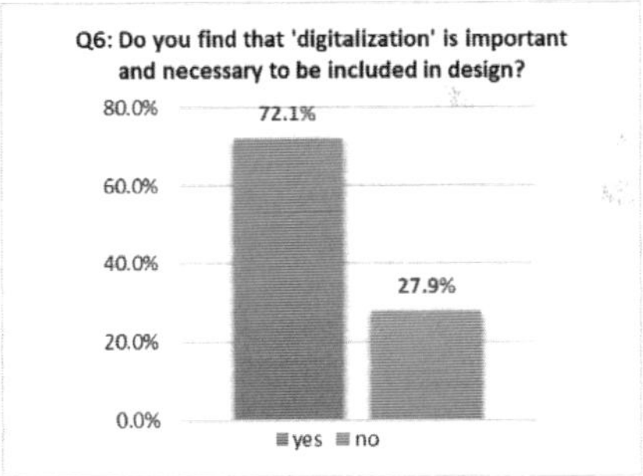

Figure 8. The result of Q6

Questions reflecting DII digitalization role

Questions Q6 & Q7 aim to find out experts' vision of the main role of digitalization, answers indicate that most participating experts, 72.1%, find that digitalization acts as the main engine in DII operation. The major progress that has happened in this aspect provides many possibilities helping designers to create new ideas that catalyze interaction with all society segments as part of the design. Furthermore, 72.1% of experts think that digitalization has the ability to change and enhance the relationships between visitors, between visitors and urban space, and finally between visitors and their surroundings, all in the same place with different levels of interaction.

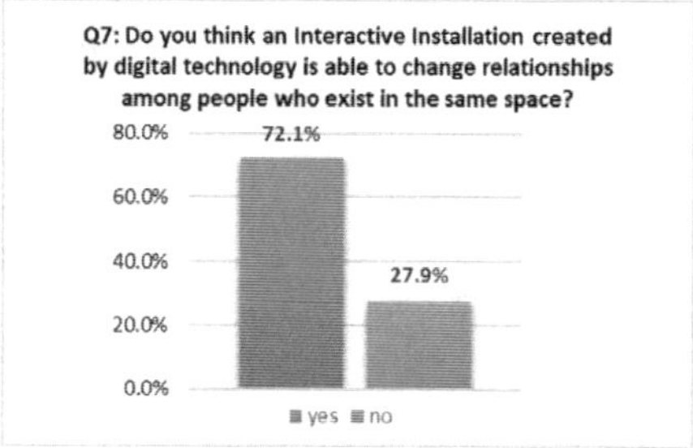

Figure 9. The result of Q7

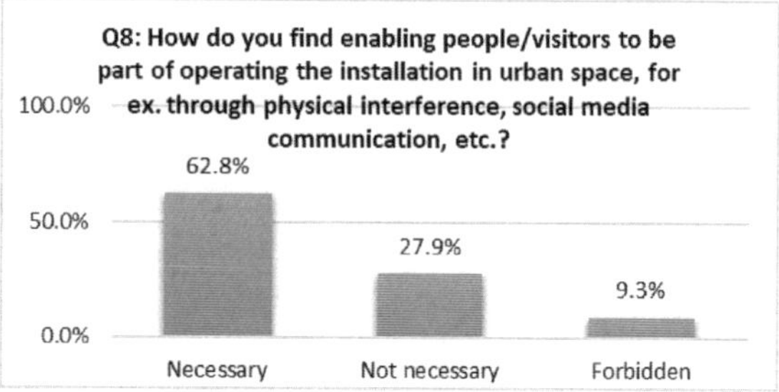

Figure 10. The result of Q8

According to the above questions, experts have been asked to support their answers with examples from Egypt. The majority thought that most city centers in Egypt are poor in urban space design and suffer from the absence of well-designed interactive installations.

Questions reflecting DII as a social interaction tool in city center

Analysis for Q9 answers show that 74.4% of the experts believe that DII can be considered one of the most attractive and interactive tools for social interaction in city centers, for they attract visitors and motivate them to interact and participate being a part of the design and operation. Some experts explain how to achieve this objective as follows:

- Through engaging private sectors and organizations.
- Increasing awareness and dedicating people to as digital staff.
- Selecting a proper geographic place within a proper social community sector that best fits to ensure the success of the application and later promotion and replication of the mode.
- Choosing touristic areas with a lot of people is the best way to accomplish that goal.
- By integrating digital visual technology and vocabulary as an essential component of urban spaces.
- User-dependent platforms must be designed to operate through their interaction with sound or motion, not just operating digitally without user interference (such as dancing fountains in many malls and public spaces).

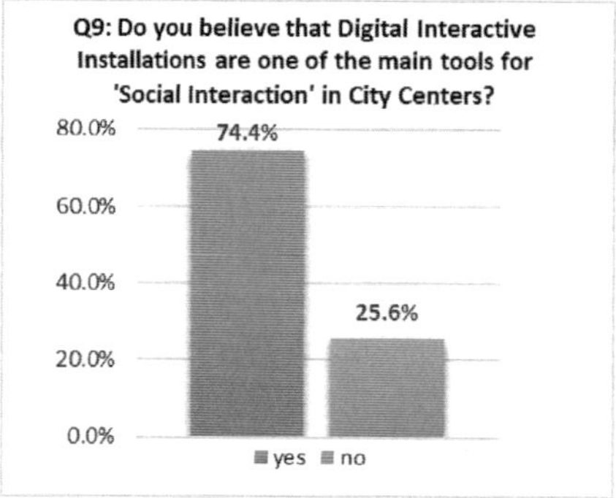

Figure 11. The result of Q9

7. Guidelines

Based on the previous review and analysis, the study was able to set specific guidelines for a successful DII design being a tool to achieve high qualities of social interaction and participation in city centers, a tool to reshape the society's urban perception engaging them with space and immersing them in new technological applications… The guidelines are presented in the following figure:

IOP Conf. Series: Earth and Environmental Science **992** (2022) 012006 doi:10.1088/1755-1315/992/1/012006

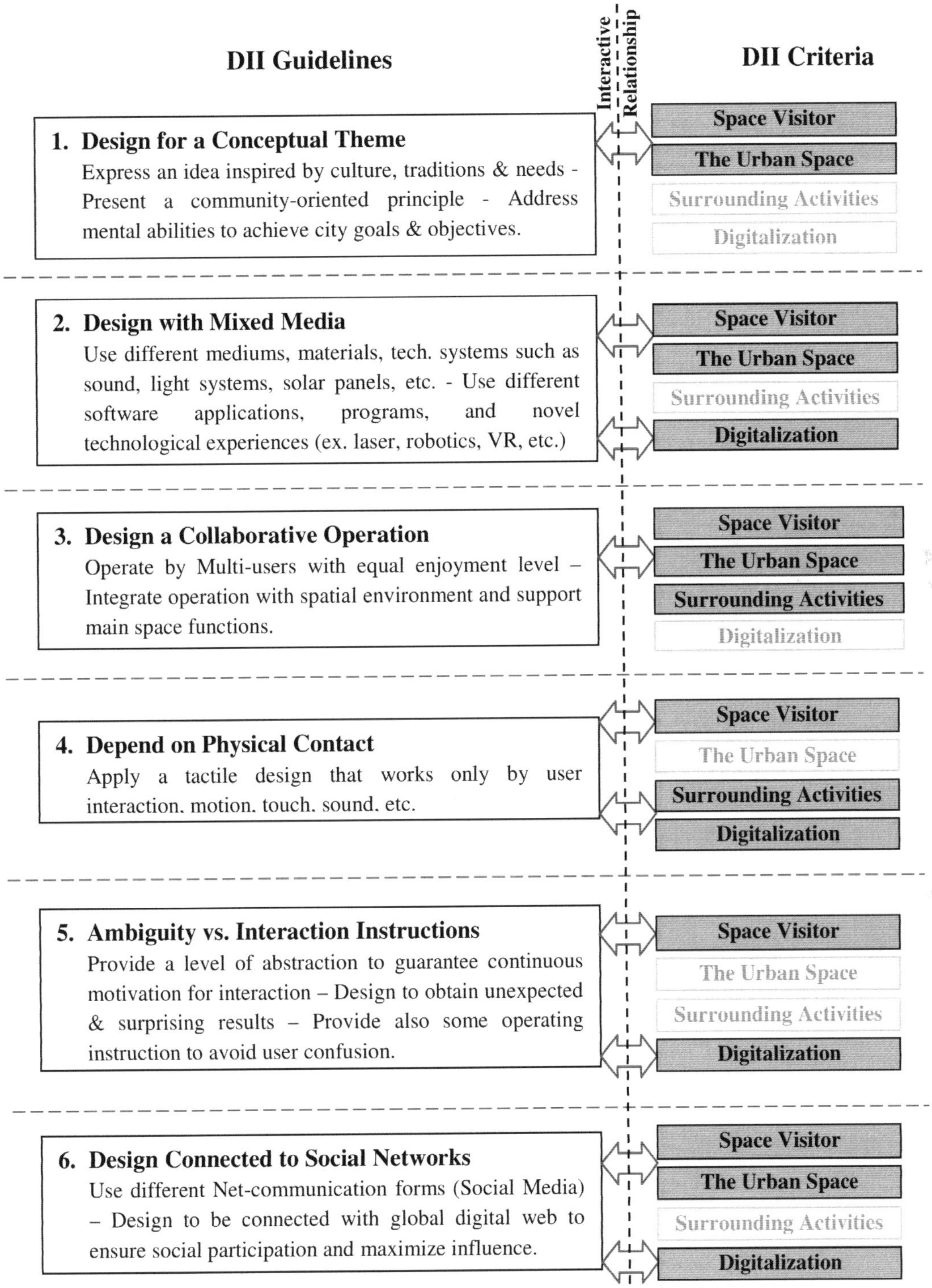

Figure 12. The proposed guidelines for DII

8. Conclusion

Digitalization, all over the world, is having great impact on cities and the way people engage with its spaces, it motivates and catalyzes social interaction and participation with space elements and the surrounding environmental context. However, still immature in Egyptian city centers. Digital Interactive Installations are one of the main tools for livable cities, more functional spaces and more community oriented experiences. Through its literature and an analytical survey, the study explores the importance and the need for DII in city centers. It sets a group of design guidelines that requires new forms of social engagement, new forms of technological interfaces and merges both in one space. It believes that DII with specific guidelines are best tools for maximizing human interaction, fostering collaboration and urban space value.

9. References

[1] Ahmed S 2018 Interaction and Interactivity: In the Context of Digital Interactive Art Installation **10** 1007/978-3-319-91244-8_20.
[2] Ahmed S, Jaccheri, Letizia, M'kadmi and Samir 2009 Sonic Onyx: Case Study of an Interactive Artwork *Arts and Technology* **30** 10 1007/978-3-642-11577-6_6 p40-47
[3] Hassan H et al 2016 Measuring people's perception towards Earth-sheltered buildings using photo-questionnaire survey *Sustainable Cities and Society* Volume **26**, Pages 76-90
[4] Jun H, Mathias F, Yu Z, and Feng W 2014 Designing Interactive Public Art Installations: New Material Therefore New Challenges *13th Int. Conf. Entertainment Computing ICEC* Sydney Australia **10** 1007/978-3-662-45212-7_25 hal-01408522 pp.199-206
[5] Ryszard W and Kluszczynski 2010 Strategies of Interactive Art *Journal of Aesthetics & Culture* **2:1**, 5525, DOI: 10.3402/jac.v2i0.5525
[6] Vliet V H 2019 the Atmosphere of Open Public Spaces
[7] Zhao S and Kim C 2021 Applications of Interactive Installations in Public Spaces *Psychology and Education* **58**(5): 2947-2952
[8] Ed Reeve, 2018 Studio Ini's Skeletal and 'Disobedient' Installation for the Greek Entry at London Design Biennale Responds to Body Movement, Available from: https://www.archipanic.com/disobedient-design-installation-in-london/, Access Date 10.2021
[9] Salah Uddin Ahmed, Letizia Jaccheri, Samir M'kadmi ,2009, Sonic Onyx: Case Study of an Interactive Artwork, Arts and Technology, Arts and Technology - First International Conference, Arts IT 2009, Yi-Lan, Taiwan, Springer Berlin Heidelberg
[10] The District Detroit, 2018, Prismatica' Art Installation Debuts in Beacon Park, Available from: https://www.districtdetroit.com/news/prismatica-art-installation-debuts-in-beacon-park, Access Date 9.2021
[11] Alexandre Lemieux, 2021, A modern ice palace, Available from: https://www.creos.io/en/portfolio/prismatica/ , Access Date 9.2021
[12] TechTarget Contributor , 2021, Communication with digital technology, Available from: https://whatis.techtarget.com/definition/digital, Access Date 10.2021
[13] Designing Buildings is the construction wiki, 2021, Digital technology, Digital infrastructure Available from: https://www.designingbuildings.co.uk/wiki/Digital_technology, Access Date 10.2021

IOP Conf. Series: Earth and Environmental Science **992** (2022) 012007 doi:10.1088/1755-1315/992/1/012007

Smart public transportation: *A future framework for sustainable new cities* (Case study Greater Cairo)

Hayam M Omayer[1]

[1]Assistant Professor, Architecture Department, Cairo Higher Institute for Engineering, Computer Science and Management, No.2, Banafseg Center2, first Settlement, Cairo, Postcode 11477, Egypt.

E-mail: hayamomair@gmail.com

Abstract. In the context of the tremendous progress in information technology in all spheres of life, it is necessary to adopt more intelligence solutions to increase the quality of life and achieve sustainable cities. Currently, without a doubt, we have the ability to make all city services accessible to everybody. Transportation and mobility systems are critical components of sustainable city development. Therefore, the scientific literature on the technological systems employed in transportation systems and their primary uses is extensive, outlining their application to improving people's quality of life through promoting a sustainable environment. Recently. There is a new generation of ITS (intelligence transportation system) resulting from the encouragement of related public policy that has been deeply integrated with new infrastructure and new technologies for improving the quality of life. The new generation of ITS technology and urban analysis and planning technology can not only meet the current problems of the coordinated development of the public transportation system and big cities like Cairo, but also meet the needs of the city's future development in advance. This paper will discuss the transformation of intelligent public transportation, the opportunities, and applications, using technologies, and implementation in various urban areas using a case study of China's smart transportation. Additionally, propose a future framework for sustainable transportation in new cities and the feasibility of implementing them in Egypt's new cities. Highlight the actions and outcomes of integrating technologies, as well as their environmental, urban, economic, and social impacts. In order to develop and increase the quality of life in the cities of Egypt and the developing world in order to achieve sustainability goals.

Keywords: Smart Transportation, sustainable city, Transport in GC, ITC

1. Introduction

In developing countries, the conventional methodology for handling the transportation congestion issue has been, for a long time, to include more open transport centres, incorporating prepared stations, quick travel stations, transport stops, etc., and constructing new ones, in this way permitting better traffic conditions, review [1][2]. This approach isn't the better solution, in terms of retaining a colossal sum of rare monetary and land resources, causing environmental intrusion, and generally increasing the environmental and safety hazards [3]. For example, greater Cairo (GC) is experiencing severe traffic congestion, which is resulting in delays, a decline in safety records, and significant environmental costs in terms of pollution and energy usage [4]. For a comprehensive examination of GC's urban transportation issues [5]. The tendency has been to adopt a significant change in thinking. While many countries' resources are becoming scarce, the proclivity has been to embrace data and communication innovation that enables the most effective use of techniques and applications. The notion of sustainability is evolving via more participation, as regional sustainability is defined as achieving a balance of social, economic, and environmental sustainability. At the city level, sustainability is determined by the extent to which cities are adaptable to changing requirements and the capacity of its elements and activities to adapt to the numerous changes in society through time [6]. For a long period

IOP Conf. Series: Earth and Environmental Science **992** (2022) 012007 doi:10.1088/1755-1315/992/1/012007

of time, countries around the world have aimed to attain sustained economic development. Sustainability displays a commitment to reducing resource and material use in order to ensure the survival of future generations. Transportation in metropolitan areas is critical for encouraging economic development on a sustained future. Therefore, in advanced countries, urban smart transportation is moving towards better electrification, automation, and connectivity rapidly. At the same time, as the market for personalized, diversified, and high-quality travel services continues to expand, the innovation process of urban public transport service models is also accelerating [3]. With online appointments, sharing, and customization, new types of bus services with characteristics are emerging. In addition, as new technologies such as big data analysis, BIM (Building Information Modeling), and GIS (Geographic Information System) are applied to the planning and construction of major cities [7] such as China, which makes it possible to improve, break through and innovate the classic theories and methods of urban public transport planning, design and operation management [8][9]. The case study and comparative study of typical cities in China and Egypt can also be used for the development of the urban public transport industry and equipment systems in China. Manufacturing and technological applications increase the value of popularization and have a radiation effect. The study was conducted as part of a series of studies targeting the integration of smart technologies that provide various solutions able to solve our challenges in Egypt, especially those related to development and increasing the quality of life in the cities of Egypt and the developing world in order to achieve sustainability goals. The series of studies was conducted in collaboration with the CHI Institute and Southwest Jiaotong University with STIFA. The research is part of the author's participation in the

2. Literature review

2.1. Public Transportation background in Egypt

According to the World Bank statistical analysis, The Great Cairo has a huge population, more than a fifth of Egypt's population [10]. Along with being one of the world's largest cities, it faces the same environmental, economic, and urban constraints as many other major cities in developing countries [4]. The Egyptian government has been building new cities to accommodate the increasing population. Most of the new cities are not covered by a public transportation system. Therefore, a lot of citizens move to their work using their own cars or informal public transportation, which leads to traffic congestion. Traffic congestion results in a lot of wasted time, which reduces the productivity of the employees, increases fuel consumption, reduces the efficiency of the vehicles, and increases air pollution. Total daily visits climbed from 5.6 million in 1971 to 21.6 million in 2001, according to the Developing Research and Technology Planning Centre and Cairo University, and are expected to exceed 30 million in 2022 (DRTPC 2009). On the other hand, the shortage of buses means it is hard for low-income citizens with no cars to decide to move to these new areas, which affects the government's efforts to relocate citizens to new districts [11]. Furthermore, there is no way for citizens to know which buses are available in their areas and their schedules without waiting for a long time and with no guarantee of finding a suitable ride [12][13]. And if the citizens find the buses, the heavy capacity inside the buses increases the probability of infections of the citizens with viruses like COVID 19. Furthermore, public transportation providers do not have strong incentives to increase buses and rides in new areas because they are concerned about the financial outcome. This study proposes a future framework for a smart transportation system in order to address the above identified issues.

2.2. New cities of GC: the challenges and opportunities

The new cities planned to provide housing and employment opportunities for the crowded population of Cairo are attractions for the population, in order to ease the pressure on the larger Cairo region and to try to obtain integrated urban communities [14]. This resulted in the establishment of New Cairo, Sheikh Zayed and 6th October, furthermore the present construction of the New Capital. Cairo's population has more than doubled in the previous 40 years and is predicted to do so again in the next 40. The government supplies the roads and infrastructure for these new communities. However, because its people work in other areas of the metropolitan region, these new projects place additional strain on the city's transportation infrastructure, increasing transportation expenses [3] as shown in figure 1. Currently, with population growth in new cities, and transportation networks and infrastructure have trailed behind demand growth, resulting in service expansion being constrained. As a means of public transportation in Egypt, bus services serve customers with a comprehensive network that spans the governorate [15]. The CTA is responsible for organizing, management, and operation, as well as for supervising twenty-two private companies that own and manage concession-line minibus services. Regardless, the network is underfunded, leading in ageing fleets, congestion, and decreased service frequency. This substandard service pales in comparison to the advantages of shared taxi services, which provide customized and flexible stops. Table 1 explains the key issues and actions by the Egyptian government for improving the transportation sector in new cities of GC. On the other hand, despite all the challenges that have been mentioned, there is potential for new cities to find and apply solutions through which they can become sustainable cities, such as the significant urban growth occurring in GC's new metropolis, The emergence and variety of urban planning and design initiatives to the degree that they need studies, Spending on development projects. Furthermore, urban design has considerable economic potential. Using researchers' potential in a city where the new cities are a vast area for integrated studies is a good idea. By specialized bodies, promotion and facilitation of all disciplines of studies.

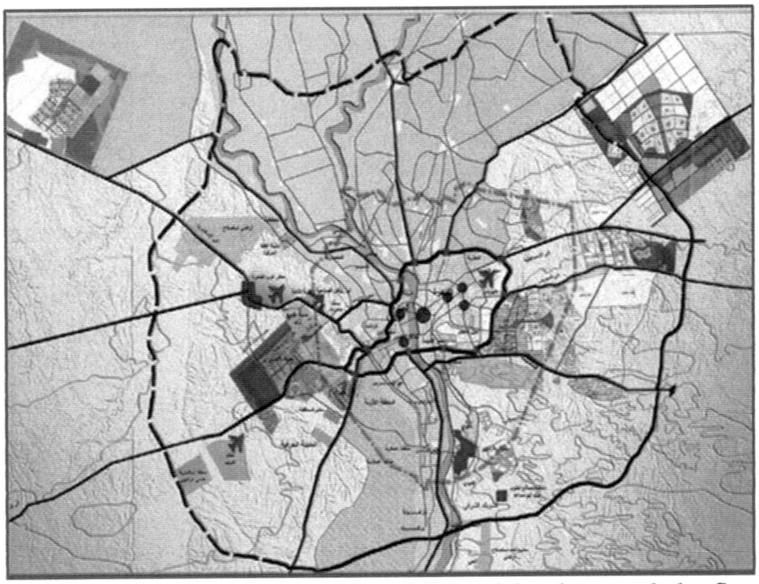

Figure.1 the relation between the new cities of GC and development hubs. Source: [14].

Table 1. Key issues and actions for transportation in new cities of GC at last 10 years.

Government challenges	Government actions at last 10 years	Reference
• Congestion exacerbated inadequate public transportation. • A huge cost in terms of higher fuel usage and commuting time lost. • High accident rate. • A huge cost in terms of higher fuel usage and commuting time lost. • Pollution of the environment (air, noise, ets.)	• Natural gas conversion of public-sector cars. • The effort to replace ageing taxicabs. • The Transportation Committee of Parliament recommended a new traffic legislation with the goal of increasing traffic safety and reducing road accidents.	• [4] • [5] • [16] • [17] • [3] • [18] • [17] • [19] • [15]

• Weaknesses and dispersion of institutions. • Financial arrangements that are insufficient. • Private cars, and their percentages are increasing. • The public transportation network does not cover subways and neighborhood centers. • The emergence of random transport such as private vehicles, motorcycles. • Adverse effects of random transport on public health and safety.	• Conducting emissions inspections as part of the vehicle licensing process • Emission reductions from motorcycles • Promoting non-motorized modes of transportation such as walking and cycling • Construction of new roads and bridges • Improving existing roads inside the cities or linking between new cities. • The government funded the purchase of new buses and the conversion of existing buses and taxis to natural gas.
Source: Prepared by author based on above references.	

All the previous problems facing the existing city and the transformations that are taking place in the extension areas point to the need for a smart transportation system capable of offering high-quality output at a lower cost, while facilitating faster decision-making, through achieving integration between existing transportation infrastructure and ICT (Information and Communications Technologies) and IOT (Internet of Things) This will greatly boost performance, improving urban transportation via improved control systems and sensor technology.

2.3. Formatting author affiliations
The sustainable city is a vision of the future. [20] Define a sustainable city as one that maximizes socioeconomic advantages for its residents while maintaining environmental and equitable standards, as assessed by suitable metrics. Environmentally conscious concepts used to describe cities have spawned projects, and cities like "green city," "eco-city," "low-carbon city," and "smart city" are all used to characterize cities that are pursuing to sustainability. According to [21], 'transportation is critical for economic, social, and environmental growth. Transportation's aim is to carry passengers and commodities effectively while having the fewest adverse effects on the environment and society' [21]. The existing transportation system is a significant contributor to air pollution, greenhouse gas emissions, and degradation of the ecosystem, climate change, and health repercussions. ITS is a critical component in resolving past issues [3]. The most effective way to meet transportation demands is through a smart transportation system. It is not an option to continue to increase the urban population and freight facing megacities and to accomplish and create sustainable cities. Numerous organizations and professionals have defined sustainable transportation, resulting in a variety of definitions. According to [3] that defined sustainable transportation as "any method of transportation that considers humanity, is economical, safe, and offers a variety of modes of travel." Additionally, it is environmentally friendly because it is based on renewable or regenerated energy rather than fossil fuels.

2.4. Concept of Intelligent Transportation System ITS
According to [22] advances in engineering and technology have led many large cities and urban areas to seek to become smart cities as a result of the integration of scientific innovations and their use to provide smart solutions and innovations to sustainable infrastructure, thus enabling them to use their resources efficiently and providing a quality of life for citizens. In smart cities, data is collected from all over the city through infrastructure containing distributed sensors and network structures with communication technology, such as traffic data, weather information, power grids, etc. Communication technology is either wireless or wireless, or both of them. [29] defined ITS in 2001 as "the intelligent and effective application of advanced technologies in areas including data, communication, control,

sensing, and system interoperation on the basis of competitive infrastructure to surface transportation in order to provide real-time, precise, and efficient transport services operating across a wide range." The intelligent transportation system is built of several sensors that perform a variety of sensory functions [22] (Tang, V. W. S., 2006). In general, intelligent transportation systems make use of sensor technology [23]. Loop detectors monitor traffic volume and accidents, automotive type and presence, and speed. To provide a more secure, effective, and timely transportation service for goods and city citizens, the transportation system is currently undergoing a transformation into ITS via V2V, V2H, and V2I communications, in addition to automobile automation and its associated communication structure. Figure 2 shows the concept of a Smart Transportation System [24].

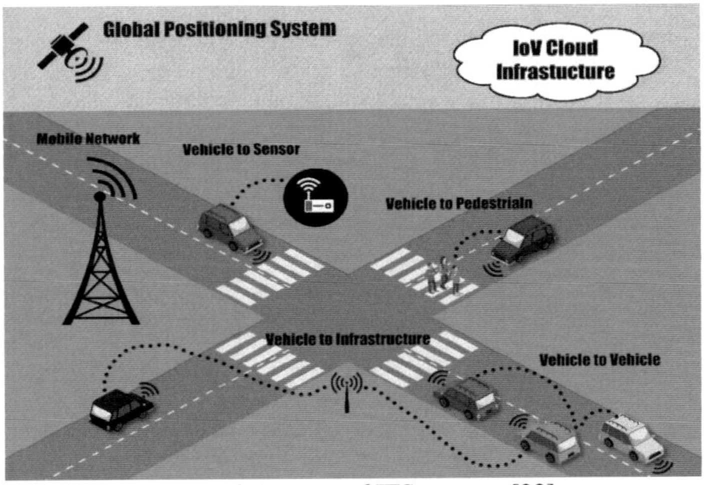

Figure.2 Concept of ITS, source: [23]

3. Research Methodology

The paper is structured as follows. Section 1 reviews the current transportation background in Egypt, the novel concept of ITC, the opportunities, and applications, using integrated technologies. And studding its environmental, urban, and economic impacts. Section 2 adopted a descriptive case study conducted on one of the largest cities with a population of more than 20 million and possesses new districts similar to the new cities in Egypt, which is "China's Xiong'an New District". These Chinese cities and their successful experiences with ITS can be used as comparative research cases for scientific and technological achievements to investigate the applicability of ITS. Finally, Section 3 describes a future framework for the development of ITC that has a big potential impact on sustainability in new cities in GC, Egypt.

4. Macro Scale Case study

4.1. China's transport plans and transferring to ITS

China has a high degree of industrial maturity in the application of ITS technology. A complete, open, and powerful domestic industrial chain in China provides a full range of ITS technologies and equipment options. From the perspective of the industrial chain of ITS, the current ITS industry chains in China are fully developed (Fig.3), including communication chips, communication modules, terminal equipment, vehicle manufacturing, software development, data and algorithm provision, and high-precision positioning. The entire industry chain has formed a certain scale of competition and cooperation in all aspects. It is estimated that the market size of China's smart transportation-related market will reach 159 billion yuan in 2023. In June 2014, the official document from the Chinese Ministry of Transport "Guidelines for the Construction of Urban Public Transportation Smart Application Demonstration

Projects" made specific requirements for the smart construction of public transport in terms of system architecture, system functions, information resources, basic conditions, standards and specifications, and operation management, including building a complete set of monitoring system for public transportation operation status, construction of urban public transportation data resource center, construction of smart dispatching platform for urban public transportation enterprises, construction of passenger travel information service platform, construction of urban public transportation industry supervision platform. In 2018, the Ministry of Transport announced that 12 cities, including Beijing, Guangzhou, and Shenzhen, were awarded as "National Public Transport City Construction Demonstration Cities" and more are in the process of such development (figure.3).

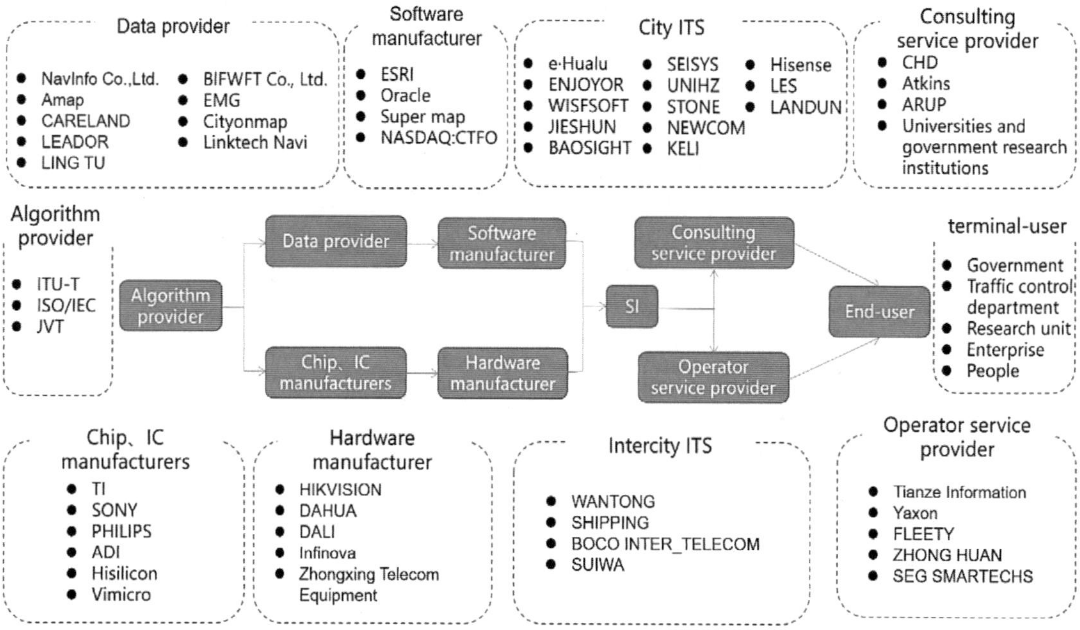

Figure.3 Map of China's Smart Transportation Industry Chain. Source: [25]

4.2. Minor Scale Case Study - Xiong'an New District

At present, there are more than 100 cities under construction of public transportation cities. The construction of public transportation smart systems has become an important development strategy for leading Chinese cities, and the creation of "transit cities" has also become the development goal of major cities. Therefore, the development and construction of smart public transport systems has a good foundation and coverage in major cities in China, and it is also the world's largest urban smart public transport trend. China's ITS construction experience can provide support for the construction of Egypt's smart public transport system. Referring to [30], China's Xiong'an New District at figure. 4 (near Beijing) is currently building the worlds most advanced ITS system to build a new urban smart transportation system with real-time perception, instant response, and smart decision-making. Promote autonomous driving demonstration applications in the Rongdong area, and build an open smart networked area [26] promote the application of shared smart vehicles, develop customized public transportation systems, and realize the smart allocation of urban public transport and shared car resources. This is a good opportunity to use this case as a reference to the new district of Cairo. Therefore, the experience of ITS and smart city applications in Chinese cities is highly adaptable and compatible with the situation in Cairo, and can meet the development needs of the new district of Cairo.

Based on city-road-station-vehicle-person " cyclic operation, a structure of ITS - smart city Synergetic system can be designed as Figure. 5 and 6.

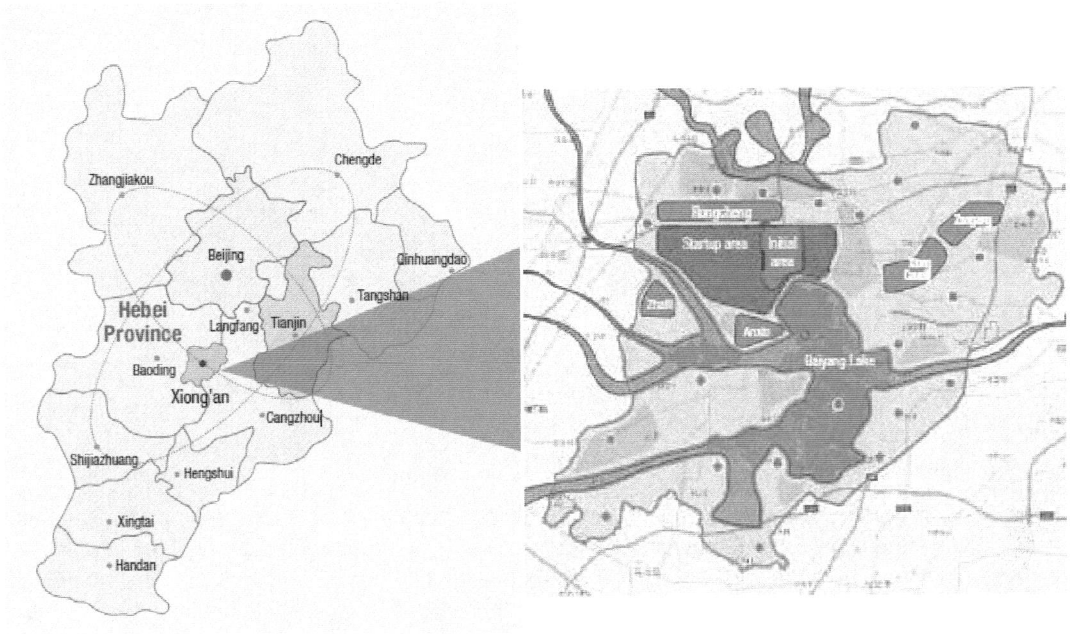

Figure.4 Map of Xiong'an New District. Source: [27].

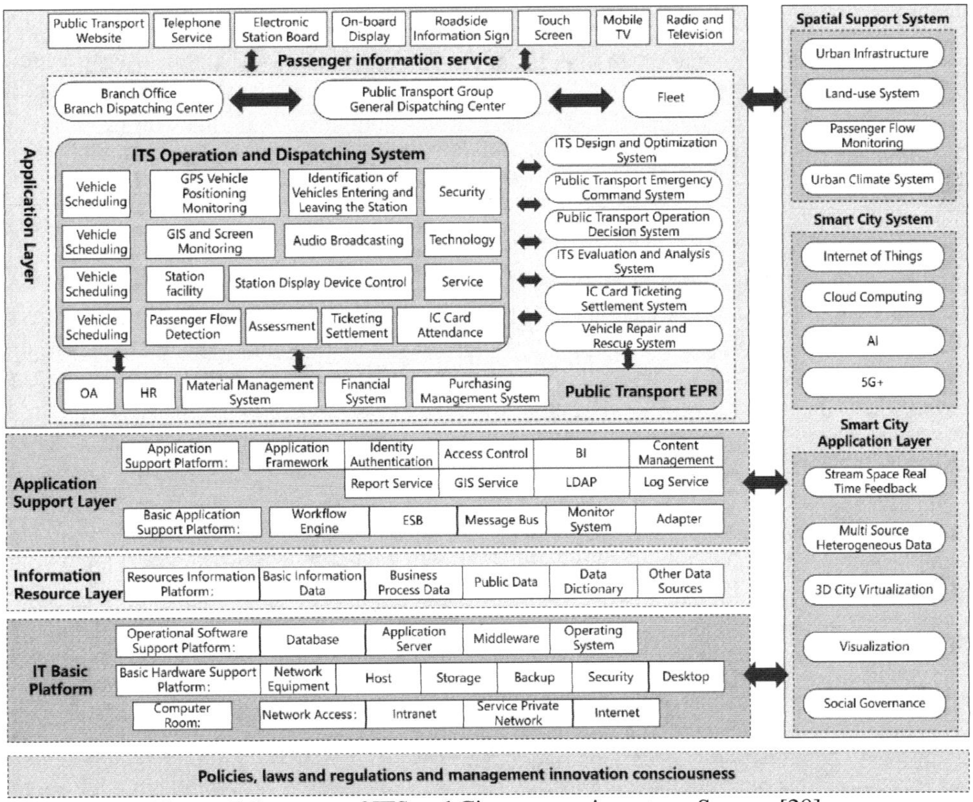

Figure.5 Structure of ITS and City synergetic system. Source: [28].

IOP Conf. Series: Earth and Environmental Science **992** (2022) 012007 doi:10.1088/1755-1315/992/1/012007

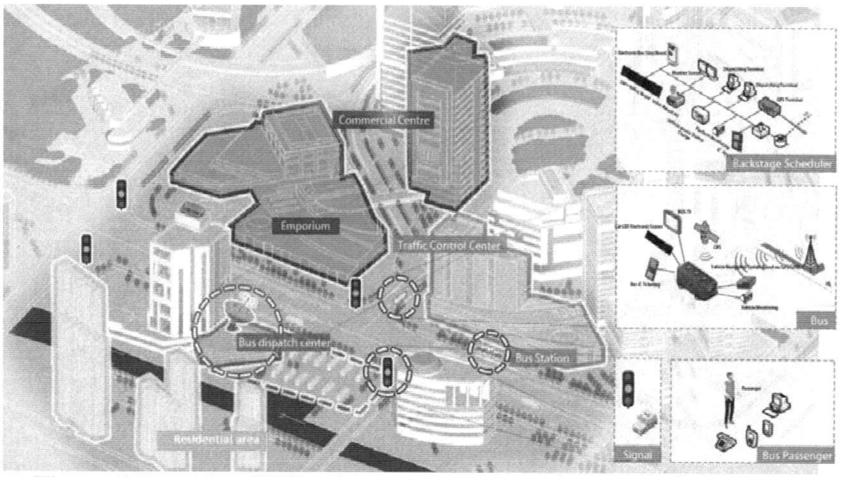

Figure.6 Scenario of ITS and smart city synergetic system. Source: [28].

5. Discussion

As followed procedures resulted and inferred from what has been studied in the previous part and the case study, the mechanisms for implement the framework will be presented in the form of procedures that illustrate each step of the proposal. Then the methodology for dealing with and the presentation of actions, outputs and impacts on the environment, urban and economic levels will be shown in table 2.

5.1. Analysis on planning elements and characteristics of public transport system for new City

Obtain the comprehensive status quo, passenger flow and traffic volume, residents' travel characteristics and needs, transportation and urban planning related professional planning of the typical city's public transportation system and the new urban area to form a basic data set for research. Analyze basic data sets, refine planning and design elements, form a public transportation system and a new urban planning element set, and deconstruct the hierarchy and structure of the element set, and use machine learning and other methods to deeply analyze the interaction mechanism between land use and urban transportation.

5.2. Feasibility evaluation of ITS in new city

Based on the interactive mechanism of land use and urban traffic, the demand database is shared, and a multi-mode network flow distribution model under dynamic information conditions is constructed. Improve and establish the application logic framework, physical deployment framework, and key technical systems (including standard specifications) of the smart public transportation system, and fully describe the blueprint of the future smart public transportation system and its service functions, and provide references for the research and development, engineering construction, and system deployment of the smart public transportation system Basis and knowledge base.

5.3. Development of smart public transport management simulation system of new city

Based on ITS standard design, with dispatching business as the main line, to meet the needs of transition from a three-level dispatch mode to a two-level mode, select typical stations and routes in the New City for smart public transportation system design and development, and build a station-level and network-Level simulation model and carry out simulation test and verification, establish the New City smart public transportation management simulation system.

5.4. ITS system Pilot Model in new city

We provide detailed information for the proposed ITS system and the required technologies for each module in the previous question. The proposed system will be applied on two buses and three bus stop stations as a starting point. Therefore, the previously mentioned technologies will be required for each bus. The proposed transportation management center is designed to have different servers to be able to analyze and store the received data from all the buses and bus stops. These servers will require licensed software as windows and SQL server and hardware as power units, rack, switches. The data management center will use an IOT platform and build a custom dashboard to present the analyzed data. Also, the data management Centre will need an API for a map such as a google map to track the buses and determine their speed and direction.

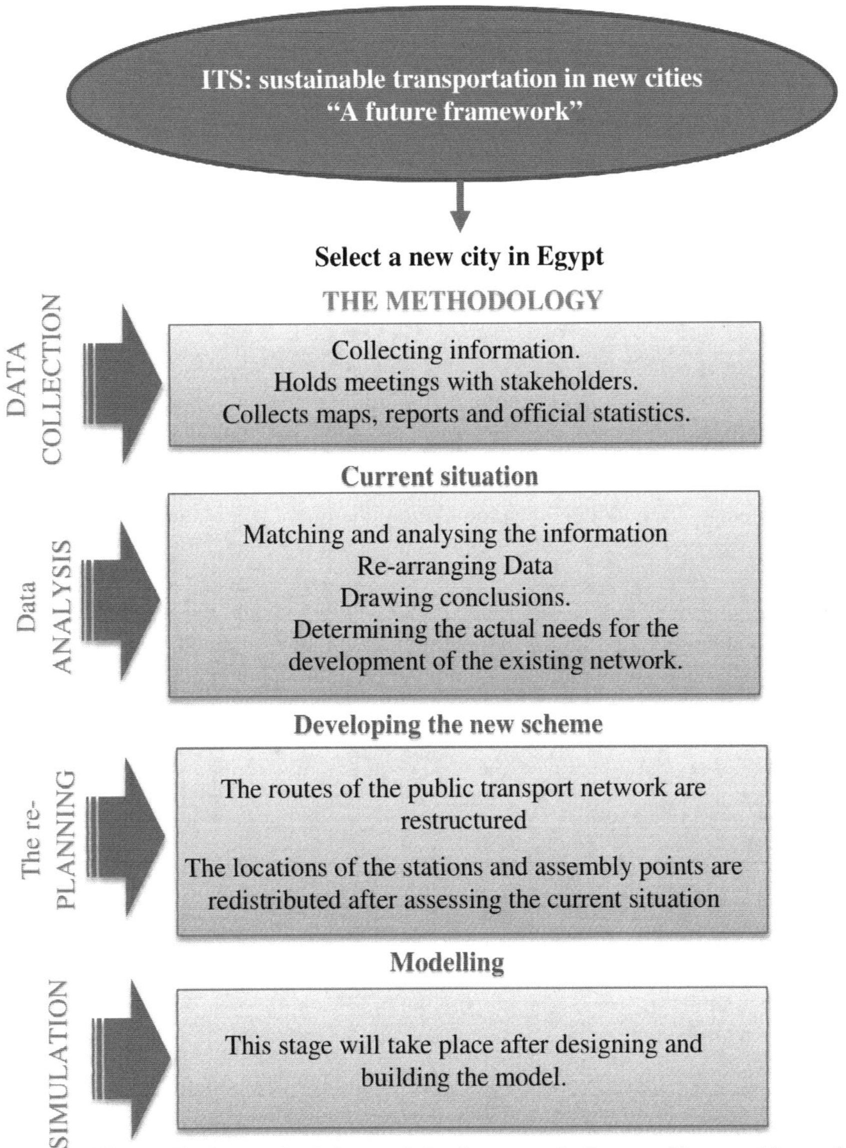

Figure.7 The methodology of the framework. Source: Prepared by author.

IOP Conf. Series: Earth and Environmental Science **992** (2022) 012007 doi:10.1088/1755-1315/992/1/012007

Table 2. The proposed Future framework for sustainable transportation in new cities

The stages	Actions	Outputs	The proposed system Impact
The design and test of the smart prototype	• Building an IOT network in the transportation sector. The design requirement is to provide travelers and authorities with updated traveling information (e.g., the arrival time for upcoming buses, bus density, and availability of seats). Provide authority with traffic monitoring and congestion information based on cameras or sensors and finally automatically saving power.	• The complete design of the IOT network will consist of IOT devices, configuration, and location of devices, gateway configuration, and programming (e.g., raspberry pi or microcontroller).	**Environmental:** • Reducing traffic congestion due to a huge number of citizens who will use this comfortable transportation network over their private car. • This will lead to reducing air pollution, wasted time and energy consumption. **Urban:** • Increase accessibility to modes of transportation by modifying interfaces and reducing cognitive demand. • Enhance real-time planning by taking into unforeseen occurrences that may benefit or detract from transportation operations. • Manage the communication and image processing part in the project (data transportation, communication, Wi-Fi).
Building the prototype	• Providing the final design equipment and devices approved and tested in the previous stage, a few transportation modes should be built as the approved design.	• This will provide tested and modified prototype. Moreover, avoid delays and any expect technical issues during the implementation.	• Make interactive transportation modes accessible to the entire population, because it serves as a public access point to a digital transportation system (DTS) that enables access to anyone without the need of applications or a smart phone.
Building a management center application	• Utilizing an IOT platform (e.g., Thingworx) to build an application to collect and analyze the data collected from IOT networks in the transportation modes.	• Showing this information to the management or authority party in map, graphs, and charts form. Furthermore, it also provides transportation modes with updated schedules and arrival times. Moreover, it alerts management, travelers, and drivers by changes due to emergencies.	**Economical:** • Improves the workability by providing the services that allow them affordable rides to work. It expands their employment options. The companies consider PT options in their selection of locating or relocating facilities. • Enable it to function as a public access point and a travel assistant for users who are low-income or disadvantaged.
the mobile application	• The mobile application for the driver will be developed in the stage of building the management application.	• The mobile application provides the driver with real-time information (e.g., route states, congestion, and expected delays). It allows drivers to report the change of route, breakdowns and to call emergency services	• Maintain an appealing and personalized user interface; this may help increase the penetration of trip planning applications and their use by various user groups (elders, immigrants, with special needs etc.). • Manage the communication and image processing part in the project (data transportation, communication, Wi-Fi).
Building the traveler mobile application	• The mobile application will happen parallel with the driver application development in the previous stage	• The traveler app will allow travelers to navigate and plan trips through the city. It will provide real-time updates to bus locations and estimated arrival time to transportation modes, the availability of seats, and transportation modes' density. Furthermore, it alerts travelers of changes of routes or schedules to their upcoming journeys.	**Social:** • Suits all the citizens in the new cities and enhances the livability of those citizens by providing them with a comfortable, unpolluted, healthy and safe place to live. • Enhance the safety and security of public transit passengers. • Streets provide access for emergency vehicles.

Source: Prepared by author.

6. Conclusion and Recommendation

The study highlights the general transportation problems in the GC, and particularly in the new Egyptian cities. The study proposed a future framework for sustainable transportation and to make it possible by implementing the smart transportation system and developing this sector. Despite that, this study faces several limitations because ITS is a vast field that consists of some groups of applications and technologies. However, these limitations can be considered for future studies and research. The research recommends that authorities and decision-makers adopt sustainable solutions for a sustainable life through emerging technologies such as IOT, big data, cloud computing, and artificial intelligence. We suggest building a synergy model of urban land planning and public transportation operation services that will form the future framework of a smart transportation system that includes a complete travel chain of sustainable "city-road-station-vehicle-person" cyclic operation. Regarding the recent achievements of "public transportation cities" and the operation experience of smart transportation systems in large and medium-sized cities in countries like China, we recommend selecting new areas of similar types in new cities with similar scale and passenger flow characteristics to the new cities of GC for comparative research and using "BIM+GIS and ICT+ IOT" as a representative technology application, the study aims to propose the system design method of smart public transportation including stations, network and management centre under the scale of urban new cities.

After designing and developing the smart public transportation system for the typical stations and routes, we constructed the station-level and network-level simulation models and carried out simulation tests and verifications. Then, considering the development conditions and needs of the Egyptian city and making a technical selection of the smart transportation facilities and equipment, we aim to build a demonstration project to operate the smart public transportation system in the new cities.

Acknowledgment

This research was conducted as part of a proposal of research project between CHI Institute and Southwest Jiaotong University with STIFA. The author would also like to express their thanks to Ahmed Haron, Abeer Farouk, and Rasha Shoitan, all of SJU, for their significant contribution to the research and applications described in this paper.

References

[1] Khisty C J 1993 *Transportation in developing countries: Obvious problems, possible solutions.* Transportation Research Record 1396: Nonmotorised Transportation Research and Issues, pp. 44-49. Transportation Research Board, Washington, D C, USA

[2] Mitric S 1994 *Urban transport strategy for Cairo: Advice and dissent. Transportation Research Record 1441: Nonmotorised Transportation around the World*, pp. 127-133. Transportation Research Board, Washington, D.C., USA

[3]Abdel Wahed Ahmed M M and Abd El Monem N 2020 *Sustainable and green transportation for better quality of life case study greater Cairo–Egypt.* HBRC Journal, 16(1), 17-37

[4] World Bank Cairo traffic congestion study executive note 2014 [Cited 2020 Jan12] Available from: http://www.worldbank.org/content/dam/Worldbank/TWBExecutive-Note-Eng.pdf

[5] Egypt SOE Report 2017 [Cited 2021 may 4] Available from: https://www.eeaa.gov.eg/en-us/mediacenter/reports/soereports.aspx

[6] Hayam M Omayer and Selim Omar 2021 *The Interaction of Building Information Modeling and Facility Management Through Sport Projects Life Cycle* 3d international conference for sustainable construction and project management – integrated management for smart cities] Cairo Egypt

[7] Ghanam R A and El-Deep A S 2021 *Upgrading urban spaces in slums as a tool to achieve Social Sustainability* (Making slums livable)-The case study of Meit-Elwan slum-Kafr El Sheikh city-Egypt.(Dept. A). MEJ. Mansoura Engineering Journal, 46(2), 67-75

[8] Al-ghamdi S A and Al-Harigi F 2015 *Rethinking Image of the City in the Information Age.* Procedia Computer Science, 65, pp. 734–743. doi: 10.1016/j.procs.2015.09.018

[9] Sodiq A et al 2019 *towards modern sustainable cities: Review of sustainability principles and trends.* Journal of Cleaner Production, 227, pp. 972–1001. Doi: 10.1016/j.jclepro.2019.04.106

[10] Ahmed Khaled Ahmed Elewa 2014 *Using the Green Infra-structure as an Economic*, Sustainable Tool for Improving Urban Life in Emerging Countries Urban Poverty Areas, Greater Cairo Case Study. European Journal of Sustainable Development 3, 4, 241-252

[11] Feisal Zeinab and Haron Ahmed 2017 *Multifunction green infrastructure towards upgrading urban ecosystem* in Cairo Conference 7ArchCairo "Interactive cities" At: Cairo, Egypt

[12] Sigalov K and König M *2016 Recognition of process patterns for BIM-based construction schedules* Advanced Engineering Informatics. Elsevier Ltd, 33, pp. 456–472. Doi: 10.1016/j.aei.2016.12.003
[13] Hamid M et al 2018 *BIM semantics for digital fabrication: A knowledge-based approach.* Automation in Construction. Elsevier 91(October 2016) pp 62–82. Doi: 10.1016/j.autcon.2018.02.031
[14] Hayam M Omayer 2017 *Human Behaviour as an Effective Tool in Urban Development at New Cities in Egypt.* Master thesis faculty of Engineering-Mattariah Helwan University. P70
[15] Padrón Nápoles V M et al 2020 *Smart bus stops as interconnected public spaces for increasing social inclusiveness and quality of life of elder users* Smart Cities 3(2), 430-443
[16] Siramy A 2018 *How choice makes city transportation smart and sustainable* [Cited 2020 Jan 12]. Available from: https://www.smartcitiesdive.com/news/ how-choice-makes-city-transportation-smart-and-sustainable/515246/
[17] Abou-Ali H and Thomas A 2011 *Regulating traffic to reduce air pollution in greater Cairo* Egypt Working Paper 664, The Economic Research Forum (ERF)
[18] Sadek G 2017 *Egypt: cabinet approves amendment to traffic law; 2017 Nov* [Cited 2020 Jan 12] Available from: http://www.loc.gov/law/foreign-news/article/ Egypt-cabinet-approves-amendment-to-traffic-law/
[19] Abdalla A 2017 *towards reforming public bus service in greater Cairo metropolitan area.* American University in Cairo, Master's thesis. AUC Knowledge Fountain. https://fount.aucegypt.edu/etds/645
[20] Chang D L et al 2018 *Knowledge-based, smart and sustainable cities: A provocation for a conceptual framework.* Journal of Open Innovation: Technology, Market, and Complexity 4(1) 5
[21] Sperling D and Salon D 2002 *Transportation in developing countries.* An overview of greenhouse gas reduction strategies. Davis: University of California
[22] Tang V W S et al 2006 *an intelligent car park management system based on wireless sensor networks*
[23] Aamir M et al 2019 *Sustainable Framework for Smart Transportation System: A Case Study of Karachi.* Wireless Pers Commun 106, 27–40 https://doi.org/10.1007/s11277-019-06259-4
[24] Javed M A et al 2016 *Security in intelligent transport systems for smart cities: from theory to practice.* Sensors, 16(6) 879
[25] Yan J et al 2020 *an evaluation system based on the self-organizing system framework of smart cities: A case study of smart transportation systems in China.* Technological Forecasting and Social Change 153 119371
[26] People's Press Editorial Committee 2018 *Guideline for Planning of Hebei Xiong'an New Area* People's Press (Sept. 2018) in Chinese
[27] Global Innovation Report 2021 *China's City of the Future: Xiong'an New Area* Hitachi Review Vol. **70**, No. **1** 036–037 [Cited 2021 sept 1] Available from: https://www.hitachi.com/rev/archive/2021/r2021_01/gir/index.html
[28] https://www.mot.gov.cn/ [Cited 2021 Dec18]
[29] Wang X 2001 *National ITS System Architecture in China* 8th ed. Sydney: World Congress on Intelligent Systems
[30] Xiong'an Green Development Research Institute 2020 *Xiong'an New Area Green Development Report (2017-2019)* – A Green Beginning of the New City China City Press (Mar. 2020) in Chinese

A Proposed Methodology for *Detecting the Urban Footprint in Egypt*

N. MAHMOUD[1], N. SAMIR[2], A. FATHI[3], W. MOHAMED[4]

[1](PhD holder), Civil Sector General Manager (AFC) - Alexandria, Egypt.
[2]Lecturer, Department of Architecture, Pharos University - Alexandria, Egypt
[3]Assistant Lecturer, Dept. of Architecture, Faculty of Engineering - Alexandria, Egypt
[4](PhD holder), Former Prof. of Architecture & Urban Design - Alexandria University, Egypt

Abstract. The detection of informal built-up environments that sprawl on agricultural land is difficult to codify specially in particular with the increase in the population in the Arab Republic of Egypt. In 2017, a law was passed to remove building infractions in rural areas according to aerial photographs taken by the military and the building codes built after the 2019 Construction Violations Reconciliation Law (CVRL) were regulated, which mandates the demolition of any buildings built without a permit that are capped at 8.2 million units Since 2007. The use of remote sensing, which is a powerful technology that relies specifically on updated satellite images, is a key tool for detecting infringements built-up regions. The European Space Agency's Sentinel Synthetic Aperture Radar (SAR) constellation is a key component of this research because of its advantages that allow for improved spatial resolution and atmospheric independence, and it takes day and night images, making it useful for a wide range of ground cover detection applications., a technological advantage that surpasses Optical sensors for the Sentinel 2 and Landsat missions that rely on clear weather conditions. The study aims to evaluate the use of radar satellites compared to optical satellites to detect urban built-up areas and encroachments on agricultural areas. The study was conducted on an area of 64 square kilometres, in Damietta City, East Delta, Egypt. Supervised classification was also used to derive the overall accuracy of radar satellite images and compare them with optical satellite images, using Error Matrix tables on Scup + QGIS programs. The results of the study resulted in reaching a high accuracy of land cover classification from the process of combining the Sentinel-1A,1B images in Maximum likelihood (ML) algorithms compared to Sentinel-2 images, reaching up to an overall accuracy (OA) of 80%, while the classification in Random Forest (RF) algorithms reached an overall accuracy (OA) of 87%. This paper recommends using the method of merging (SAR) Sentinel 1A,1B images and applying it, as well as conducting periodic monitoring of urban expansions using remote sensing techniques, to make sustainable decisions for the future of urban expansion master plans in the region and to quickly explore informal built-up encroachments.

Keywords: Urban Environment; Urban Planning; Sentinel; SAR; Urban Footprint; Land Cover; Remote Sensing Built-Up Regions

1. Introduction

Earth observation and space technology are a United Nations-recognized approach to achieving the Sustainable Development Goals (SDG) (UN, 2020). However, accurate urban mapping remains a challenge, owing to spatial, floating, and temporal constraints on remote sensing systems that often fail to obtain a rapid response (C. Small, 2018).

Recent studies have used the thermal infrared properties of urban areas and surface temperature, as well as night-time light satellite images to classify the urban footprint (Bailang Yu, 2014). The most common approach to urban monitoring applications involves the use of optical satellite sensors, as in Sentinel-2 and Landsat-8 satellites (Martino Pesaresi et al, 2016).in addition ,other studies used a combined images generated from synthetic radar aperture satellites (SAR) and optical satellites (interferometry approach), as in the study of (Castro Gómez, 2017) that had proven to generate an accurate method in urban footprint classifications and detection, where he combined Sentinel-1A (SAR) satellite images with the Sentinel 2 (MSI) optical sensor reaching a total overall accuracies of up to a of 84% for urban footprints detection. Others used the interferometric Synthetic Aperture Radar (InSAR) an approach to detect urban footprint, as it is a radar technique based on the use of more than one synthetic aperture radar (SAR) image to generate maps of the Earth's surface as well as to detect digital altitudes using variations in the wave phase of the satellite (Tomás, 2014). The (InSAR) approach is frequently used to detect landcover slides, earthquakes, and vegetation crops, as the coherence of the interferometric pair is a valuable parameter for a range of thematic mapping applications. Nevertheless, this approach is not frequently used to detect urban footprint.

However, the main advantages of radar sensing data are improved spatial resolution and high revisit frequency, and the capture of microwave characteristics of the urban environment (interferometric coherence, speckle divergence and backscattering phenomena), that make them useful for a wide range of land cover detection applications over the optical sensors which depends on clear weather situations. (SAR) satellites the Sentinel-1A/1B constellation acquires data in single or dual polarization with a revisit time of 6 days. The 6-day repeat cycle provides global coverage and enhanced performance for (InSAR) applications (Wegmüller U et.al,, 2015). Sentinel-1 Level 1 data are distributed by the Copernicus Open Access Hub under two product types: Ground Range Detected (GRD) and Single Look Complex (SLC). The pre-processing of the (GRD) products is simpler, less time consuming than the (SLC) data (ESA, 2021). As the GRD dataset comprises (VV and VH polarization channels) acquired during ascending and descending orbits over the Earth's surface. The study of (Semenzato et al., 2020) (InSAR) detected urban footprint reaching 85% values ranging from to 90% of overall accuracy of urban footprints. The pre-processing workflow of the Single look Complex (SLC) of S1A images was generated through 9 steps in (SNAP) ESA platform, unsupervised classification was applied, and accuracy assessments were performed by means of the Semiautomatic Classification Plugin (SCP) implemented in QGIS. Also, the study of (Kumar, 2021) uses (InSAR) approach on C-band but with (GRD) datasets acquired from Sentinel-1A/B sensor, and the Google Earth datasets for ground truth verification to detect urban footprints. Where, the pre-processing workflow of the acquired data was generated through 6 steps in (SNAP) toolbox platform, as it was observed that the Refined-Lee filter performed well to provide detailed information about the various urban objects. The experimental results revealed that the proposed technique works well over traditional methods in terms of accuracy and precision. However, the study didn't provide an accuracy assessment step to evaluate the overall accuracy measurements.

The selected method in this study is based on exploring the local area statistics (speckle divergence, backscattering coefficient sigma0) channels extracted from the pre-processing of satellite image of Sentinel 1(GRD),and focusing on the Sentinel 1B capabilities, as the study aims for (a) Exploring the advantages of combining the Sentinel 1B sigma0, and the capabilities of its (VV,VH) polarization channels to enhance the process of (InSAR) urban footprint detection, (b) Simulation and evaluation of the overall accuracy of built-up area classifications of the proposed methodology. The presented workflow offers a benchmark for the development of (InSAR) technique to detect urban areas, generating a reliable information of interest to a wide range of communities.

Results from this study should be used to improve existing and develop new urban area detection methods based on Sentinel-1B data.

IOP Conf. Series: Earth and Environmental Science **992** (2022) 012008 doi:10.1088/1755-1315/992/1/012008

2. Structure

To promote the use of satellite virtual constellations by means of data fusion techniques, A standard generic pre-process workflow to Copernicus Sentinel-1 GRD data is here presented. As The workflow aims to apply a series of standard corrections to apply precise orbit of acquisition, perform radiometric calibration, apply terrain correction. Additionally, the workflow allows to stack (GRD) Sentinel-1A products to Sentinel-1B data grids. Figure 1 represents the flow chart of the methodology steps.
The methodology steps as follow:

- First, validating process for Sentinel 2 image in the region of interest
- Pre-processing of Sentinel 1A,1B and the combined image
- Supervised classification of (ML) algorithms of (LCS) on Qgis Scp-plugin (Qgis, 2020), Random Forest classification algorithms (RF) classifications are applied to satellite images using SNAP TBx.
- Evaluation of the results of classifications (ML), (RF) to extract the best overall accuracy.

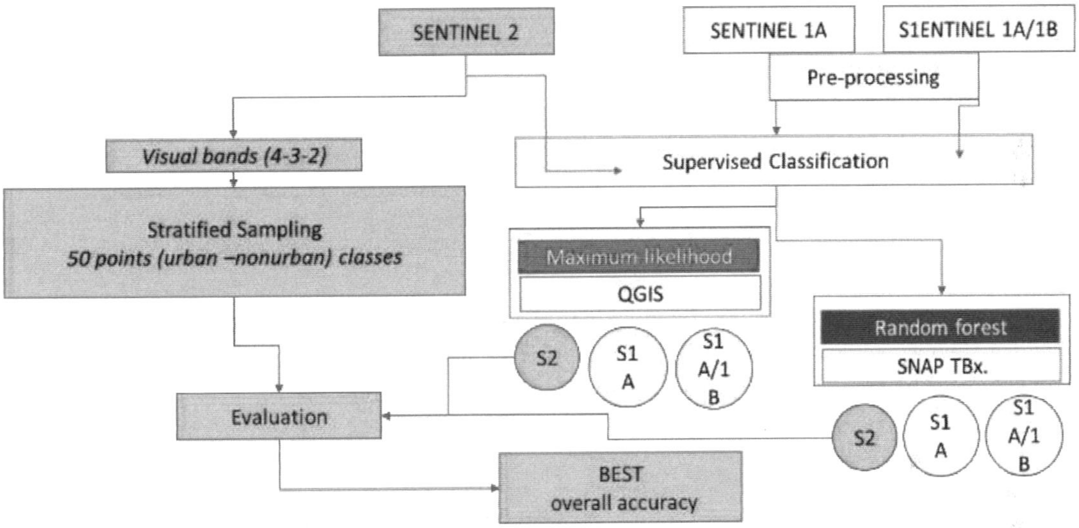

Figure 1: Methodology Flow, source: (Researcher,2020)

2.1. Study Area

The study area extends over 64 m^2 in Damietta city, which is located in the North of Egypt, bordered by the Mediterranean Sea to the North (Lon, 31 ° 45'0.00 "E, lat., 31 ° 26'60.00" N) (see Fig. 2). It contains the highest standard of living per capita compared to the rest of the governorates of Egypt also there is no unemployment due to the economic activity in this Governorate is one of the best economic patterns as it is characterized by diversity and based on the human element. This economic activity directly affects social characteristics and creates different urban environments. Damietta Port has 39 projects, 14 private free zones and 25 public free zones and is a residential attraction for the availability of free investment zones. The main industrial activities in the complex are Mopco fertilizer manufacturing and methanol production plants by Methanex (Methanex Corporation , 2015). In addition, it contains the largest port in the Middle East.in addition, it has a population of over 1 million, Köppen-Geiger climate classification system classifies its climate as hot desert (BWh), but blowing winds from the Mediterranean Sea greatly moderate the temperatures, typical to the Egypt's north coast, making its summers moderately hot and humid while its winters mild and moderately wet when sleet and hail are also common, the average annual temperature is 21.1 °C | 69.9 °F. The annual rainfall is 145 mm to 5.7 inch (Climate-data, 2021). The region has activities for extension and encroachment on agricultural land in informal urbanization activity beside industrial areas.

Figure 2: Research location, Region of interest in yellow color, Source: (Google earth,2020)

2.2. Data collection

Table 1 represents the information of satellite products and the number of bands used to derive the classification evaluations results.

Table 1: Satellites data, Source: (ESA,2020)

Satellite	Data packages	Bands
Sentinel 1A	**S1A_IW_GRDH_1SDV_20180930T155705_20180930T155730_023930_029CE0_78F7**	**4**
Sentinel 1B	**S1B_IW_GRDH_1SDV_20181014T035133_20181014T035158_013143_01848F_E9E2**	**4**
Sentinel 2	**S2A_MSIL1C_20180906T082601_N0206_R021_T36RUV_20180906T104215**	**13**

All satellite images, as well as additional data sets, referred to as the Global Geodesic System and selected projection is a global transverse Mercator in Area 33N (WGS 84, UTM Zone 33N), EPSG 32632. Areas of interest are represented in Figure 1 with a border shape file of dimensions of 8*8 km (yellow square).

2.3. Validating S2 images

Sentinel 2 was selected to calibrate satellite classifications, with a multispectral image of this satellite containing 13 strips at an altitude of 786 kilometers (ESA, 2014), these capabilities enable us to distinguish between urban and other areas using sets of false colors of the satellite image bands B, G, R (4.3.2). In this way real ground points already known in the region can be allocated. 50 points for each class of urban and non-urban land cover category are set on the visual image to assess the classification stages on confusion matrix tables.

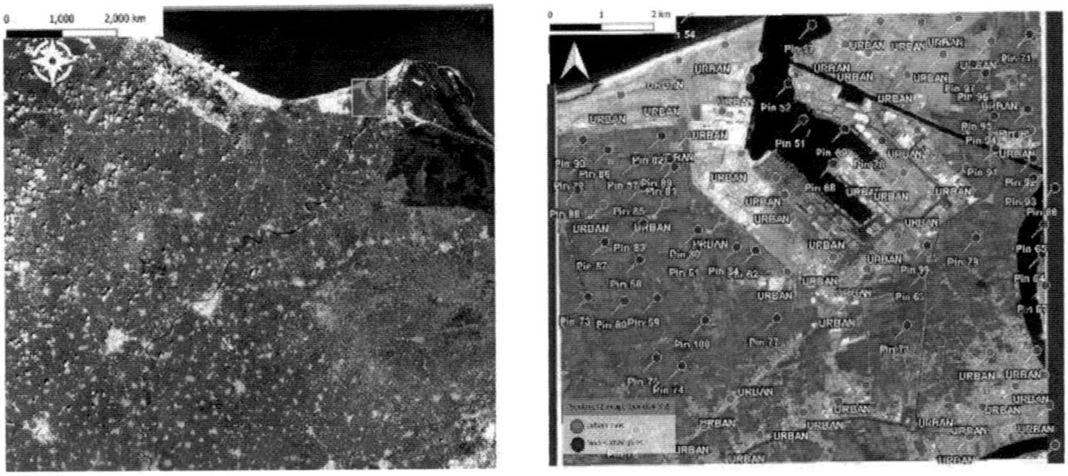

Figure **3**: Stratified random sampling of S2 RGB composite 2,3,4 bands, (a) S2 original product *(ROI Represented by the magenta color rectangle,12 bands of s2).* (b) S2 RGB composite (*false color* 4-3-2 bands), Source: Researcher working on SNAP TBx, (ESA, 2020)

2.4. Pre-processing S1A, S1A/1B

The workflow of the Sentinel 1 pre-processing stage, on SNAP TBx, illustrated in the following diagram (Fig. 4), and summarized in the following steps:

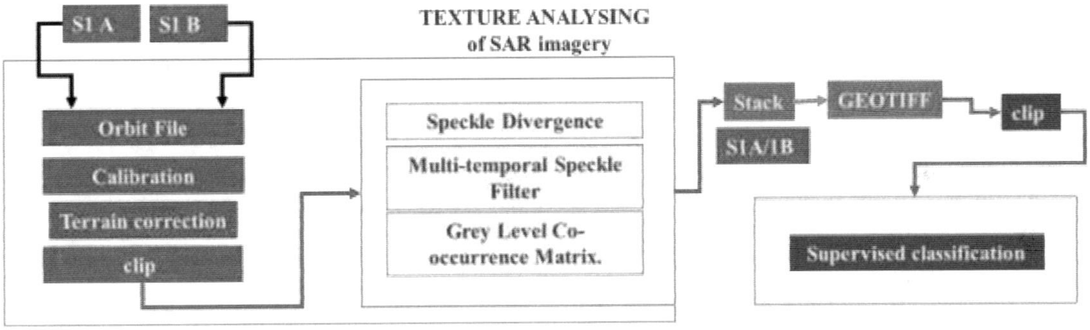

Figure 4: Sentinel 1 pre-processing workflow, Source: The Researcher

The steps of pre-processing of S1A/1B as follows:

- Pre-processing sentinel 1A and 1B images begins with the updating of orbit metadata to provide accurate information about the location and speed of satellites. Sentinel-1 Ground-Range Detected (GRD) products are geographically encoded, and geocoding ensures that the image is not only displayed in the correct location on the earth's surface, but also displays the image as expected when viewed outside a spatially enabled framework.
- Followed by the Radiometric calibration – where Values of the acquired satellite images are transformed to represent reflectivity measures of the ground surface as to calculate sigma0 of both the amplitude and intensity polarization channels.
- Terrain correction, to flip the images automatically and to add the DEM files on the stack composition, Speckle filtering –using the lee sigma algorithms reduces the variance of the complex scatterings.
- SAR Image texture Analysis (GLCM) to enhance urban extraction by displaying the back-scattering coefficient (sigma0) for VH and VV polarization, which is the pattern of differences

in image density and represents a valuable tool in improving the accuracy of the land cover classification (MathWorks, 2020)

The intensity of the Sentinel 1A image for VV-VH polarizations' bands is shown in the (Fig.5. a, b). The maximum intensity for Sentinel 1A image VV polarization reaches (14252.190 W/m^2), while the VH channels maximum limits of intensity reaches (1297.811 W/m^2). The levels of high backscattered intensities on elevated topographies and built-up environment in the region of interest are represented by the white toned pixels in the images. The Sentinel 1B image was radiometrically calibrated and terrain corrected to project the image to the location of the study area, the intensities values of the 2 channels shows high values specially in the VH band of the S1B (shown in the histogram for the polarization channels) (see Fig.5 c, d).

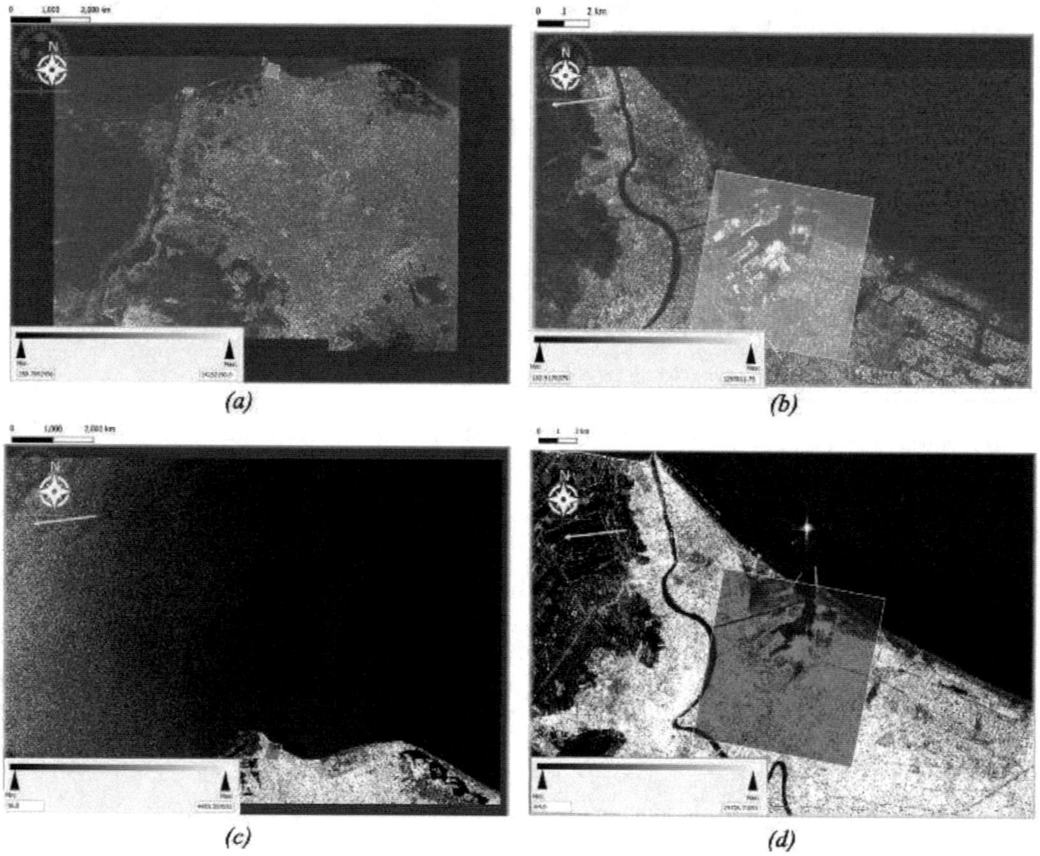

Figure 5: Sentinel 1 pre-processing workflow and channel intensities, (a) S1A (VV) , (b) S1A (VH), (c) S1B (VV channel), (d) S1B (VH channel) , , Source: Researcher working on SNAP TBx, (ESA, 2021) , *ROI is presented by the yellow rectangle in S1A, by the blue rectangle in S1B, yellow arrow is the direction of sensing.*

2.5. Classification scenarios

Two types of supervised classification have been used to classify images of S1A, S1B and S2 with the same form training shapefile in Figure 6, where it was classified in 2 classes (urban, nonurban). The first type of classification is the maximum likelihood application (ML) using the semi-automated plug-in (Scp) on QGIS. the second type, Random Forest (RF), is classified under the supervision of satellite imagery using TBx SNAP, to apply further spatial analysis.

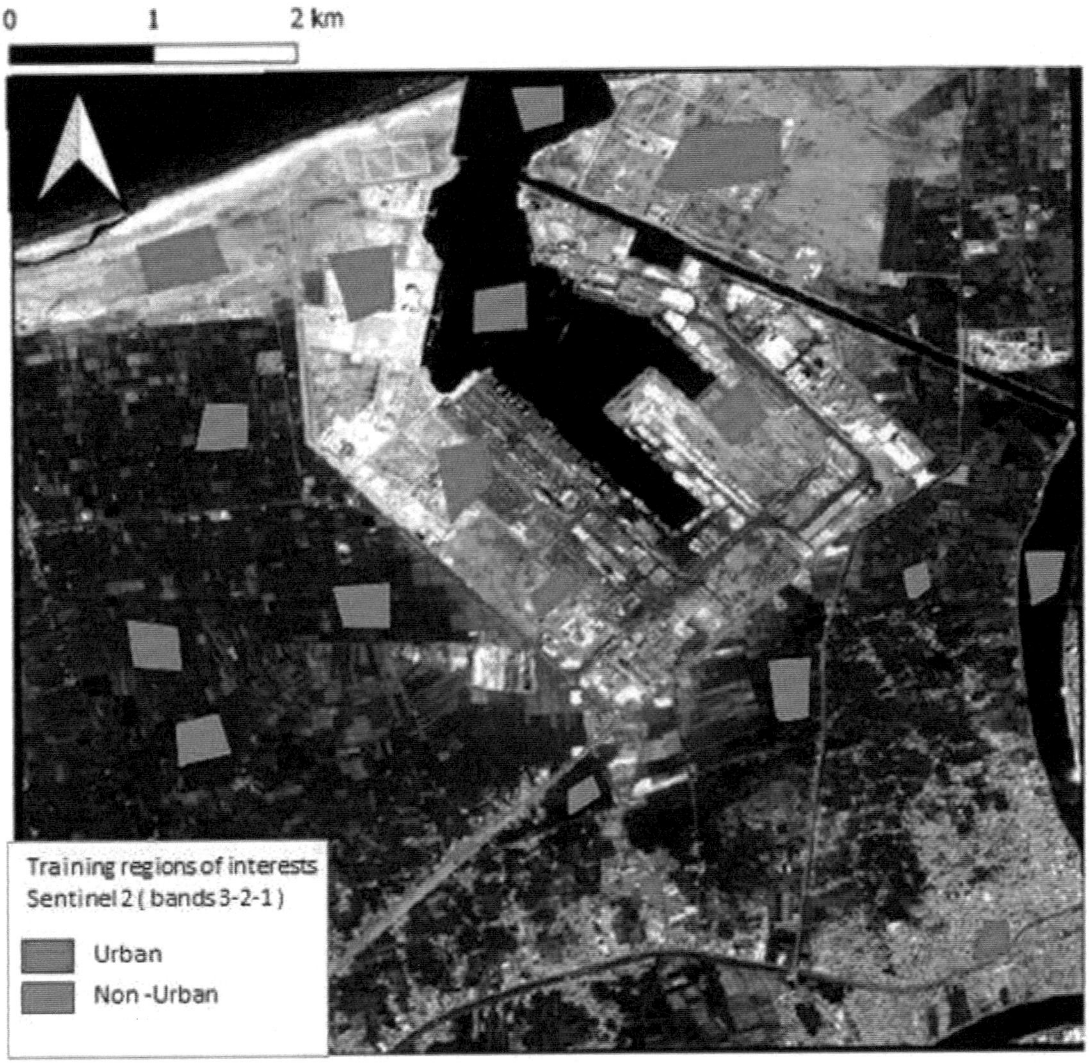

Figure 6: Training ROIs of satellite images, Source: Researcher working on QGIS, 2021

2.6. Accuracy assessment

The automated Error Matrix (also known as the confusion matrix, correlation matrix, or covariance matrix) has been used to evaluate the results of the satellite images classification, and this approach represents one of the exact methods of extraction in many statistical measures, calculating the overall accuracy, percentage of commission, omission errors and Kappa coefficient (K) of the classification of satellite images (G.Congalton, 1991).

The accuracy assessment is a measure of consistency with the information documented in a spatial information's, and is used to assess the quality of maps developed from remotely sensed data (Hashemian et.al,, 2001), using stratified random sampling, which is to divide the area of interest into smaller areas according to the category of maps, where a number of categories of land cover maps are selected and 50 points are randomly assigned to each category of land cover (Cochran, 1977).

3. Results

3.1. Pre-processing results

The Sentinel 1 satellite image pre-processing added more accurate information's for the detection process of urban footprint, as the workflow started with the orbit file operator and the calibration process, the (sigma^0) was calculated for each (VV, VH) polarization channel (see Fig.7). The S1A satellite image offers a higher set of backscattering values for the VV channel than VH polarization values, (high density values represented by white pixels in Fig.7 a). Visual image monitoring showed high intensities values of sigma^0 for pre-processing of the S1B image (Fig.7. c, d). Built-up areas on Sentinel 1B image can be easily distinguished in the VH polarization range, where the difference between high luminance of built-up areas and low luminance from plant and water areas can be detected significantly (Fig.7. d).

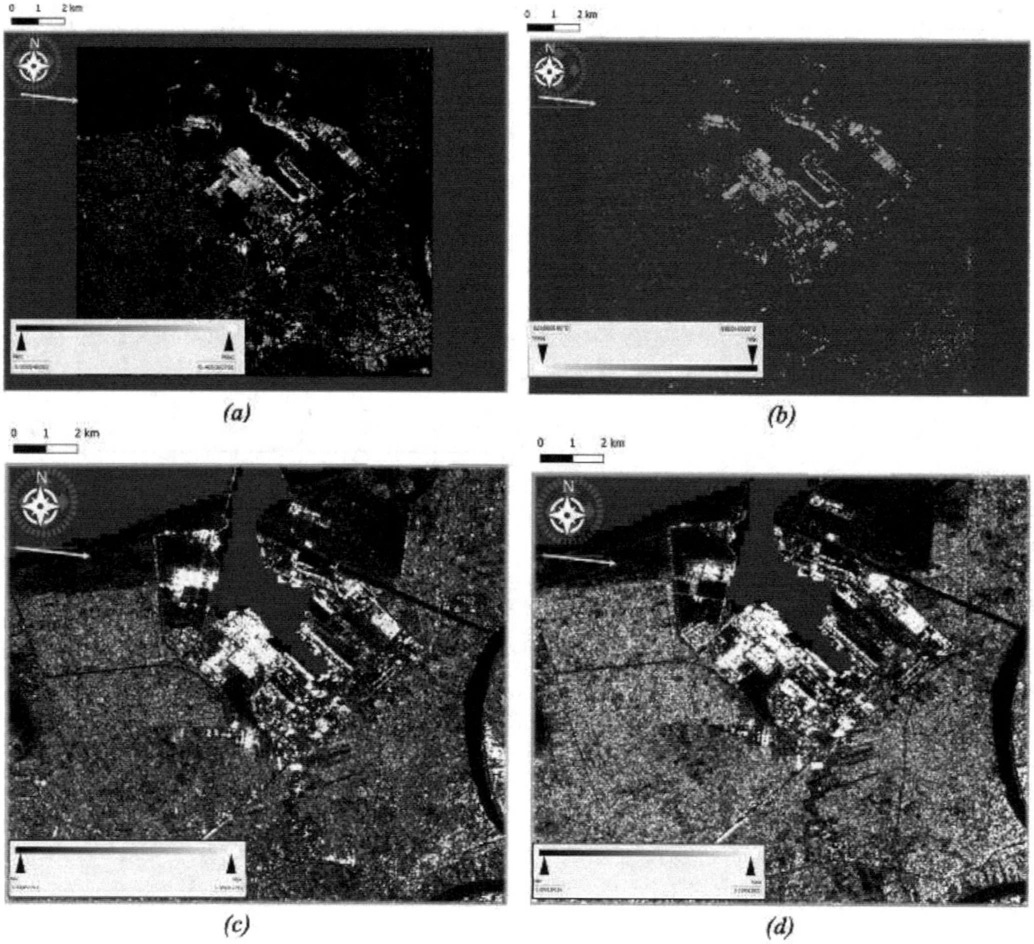

Figure 7: Processed Sentinel-1A, 1B product (sigma^0 values) t, (a) S1A-VV polarization, (b) S1A-VH polarization, (c) S1B- VV polarization channel, (d) S1B- VH polarization channel, Source: (The researcher processing sentinel 1A on SNAP TBx.,2021)

The process of extracting Speckle divergence band in Fig.8. a, and GLCM Variance of the processed image was generated by stacking both channels of the amplitude and the intensity of S1A,1B satellite images (Fig.8. b). The process maximized the intensities of the backscattering channel reflected from the built-up regions (represented in white pixels in Fig.8. a). However, in the process of extracting the GLCM, the small differences in grey color pixels distinguished the urban areas and the small regions of urban settlements.

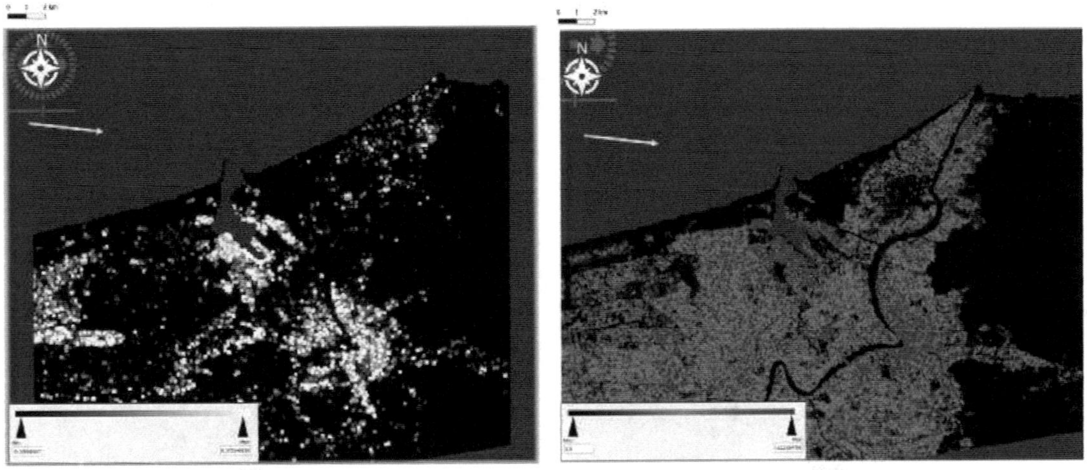

(a) (b)

Figure 8: Sentinel 1 A/B Texture analysis process, (a) Speckle divergence band, *(b) GLCM band,* Source: (The Researcher processing Sentinel 1A/B on SNAP TBx.,2021)

3.2. Classification results

In the supervised classification of the Sentinel-2 satellite image (Fig.9), Random Forest classification of Sentinel 2 represented more information's of urban pixels count due to the sophisticated algorithm that capture small details for the building's shadows and scattered pixels around the water surfaces (Fig.9. b).

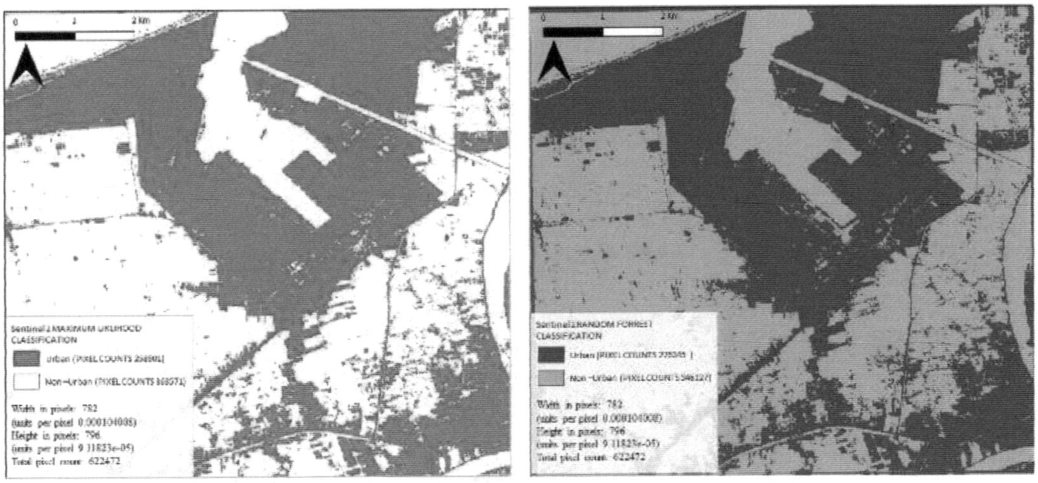

(a) (b)

Figure 9: Supervised classification of S2 image, (a) Maximum likelihood classification of S2, (b) Random Forrest classification of S2., Source: Researcher working on QGIS, 2021

Figure 10 represents the results of the classifications scenarios for S1A and the combination method of S1A/1B, where the images was classified by the Maximum likelihood (ML) on (Scp-plugin) Qgis interface (Fig.10. arc), and then classified with Random Forest (RF) classification algorithm on SNAPTBx (Fig.10. b, d), Table 2 show the pixel counts of each class (urban, non-urban) in satellite images.

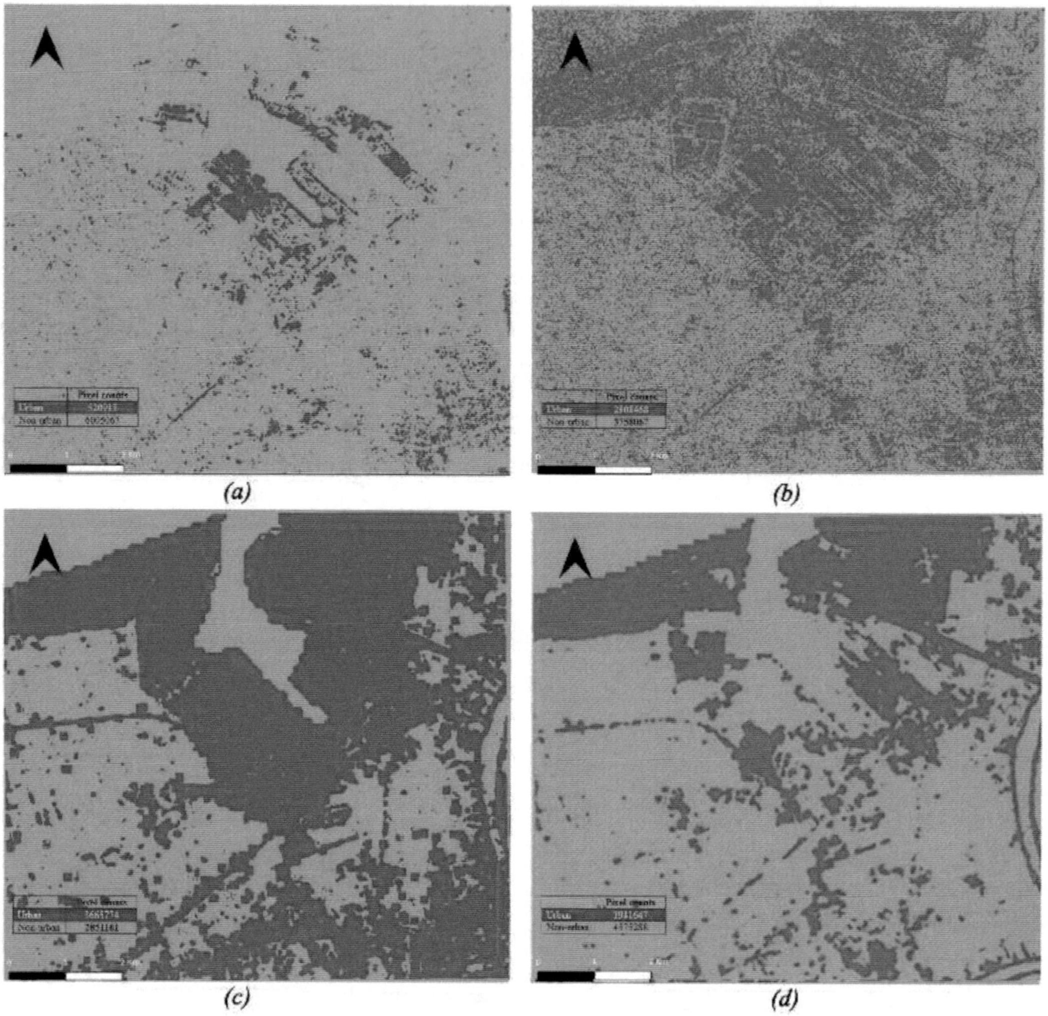

Figure 10: Sentinel 1A Supervised classification, (a) S1A -(ML) classification, (b) S1A -(RF) classification, (c) S1A/1B -(ML) classification, (d) S1A/1B -(RF) classification. Source: Researcher working on QGIS, 2021

Table 2: Classifications (pixel counts)

ML-Classification	**S2**	**S1A**	**S1A/B**
Urban	2358588	520913	3665774
Non-urban	3312131	6005067	2851161
RF-Classification			
Urban	2514739	2808468	1941647
Non-urban	3149755	3758067	4575288
	Datum		

3.3. Evaluation & Accuracy assessment results

The Scp plugin automates the generation of Error matrix tables and calculates the overall accuracies of (ML) classifications. The confusion matrix tables produced an (OA) (60%) for (ML) of S1A classification in Table 3, and (OA) (80%) for S1A-1B in Table 4, based on calculating the errors between four categories (True urban, False non-urban, False urban, and True non-urban).

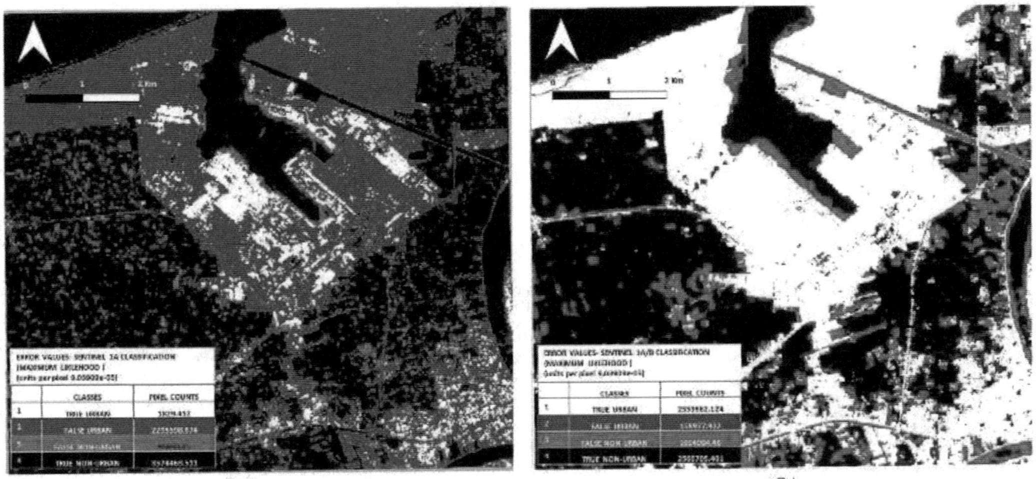

(a) (b)

Figure 11: Evaluation of (ML) classification on Scp-Qgis referenced to S2 classification, (a) Sentinel 1A pixel errors, (b) Sentinel 1A/B pixel errors. Source: Researcher working on QGIS, 2021

Table 3: Sentinel 1A (ML) confusion matrix.

		REFERENCE S2		
		URBAN	NON-URBAN	TOTAL
	URBAN	1829.412	2295598.674	2297428
CLASSIFICATION S1A	NON-URBAN	95853.45	3574463.511	3670317
	TOTAL	97682.86	5870062	5967745
TOTAL OVERALL ACCURACY				0.5997

Table 4: Sentinel 1A/B (ML) confusion matrix.

		REFERENCE S2		
		URBAN	NON-URBAN	TOTAL
	URBAN	2553662.124	1094984.46	3648647
CLASSIFICATION S1A/B	NON-URBAN	158977.932	2566705.401	2725683
	TOTAL	2712640	3661690	6374330
TOTAL OVERALL ACCURACY				0.8032

Figure 12 represent the stratified random sampling for each classification, where 50 points have been assigned randomly on the Qgis platform. For each landcover class on the Sentinel 2 image, followed by acquiring the values of each point. where 43 points represent the urban class, and 57 points indicates the non-urban class category. To calculate the errors on the confusion matrix table, the values of the points have been matched with the other classified images of the Sentinel 1A and the combination of Sentinel 1A-1B classifications.

The overall accuracies for the classification whereas calculated where it showed an (OA) (77%) for S1A (RF) classification, (OA) (57%) for S1A-1B (RF) classification, and (OA) (87%) for S1A-1B (ML) classification represented in TABLE 5,6,7 respectively.

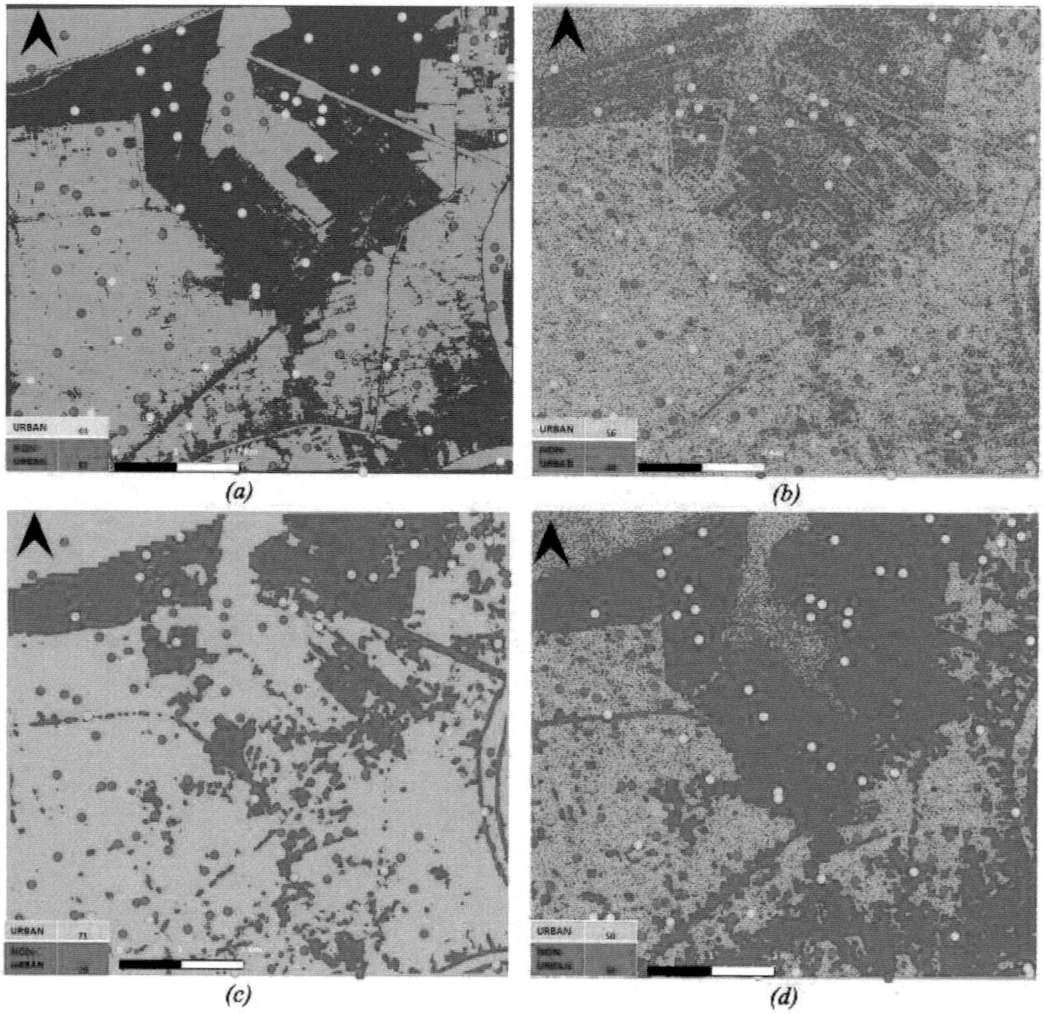

Figure 12: Stratified sampling points for (S2, S1A, S1A-1B) classifications, (a) Sentinel 2 (RF) classification, (b) Sentinel 1A (RF), (c) Sentinel 1A/1B (RF), (d) Sentinel 1A/1B (ML).

Table 5: Confusion matrix, Sentinel 1A (RF) classification.

	REFFERENCED IMAGE (S2)			
CLASSIFICATION S1A RF		URBAN	NON-URBAN	TOTAL
	URBAN	43	13	56
	NON-URBAN	13	44	57
	TOTAL	56	57	113
Total Overall accuracy				0.7666

Table 6: Confusion matrix Sentinel 1A-1B (RF) classification, Source (Researcher,2020)

	REFFERENCED IMAGE (S2)			
CLASSIFICATION S1AB RF		URBAN	NON-URBAN	TOTAL
	URBAN	43	28	71
	NON-URBAN	28	29	57
	TOTAL	71	57	128
Total Overall accuracy				0.566

Table 7: Confusion matrix, Sentinel 1A-1B (ML) classification.

	REFFERENCED IMAGE (S2)			
CLASSIFICATION S1AB (ML)		URBAN	NON-URBAN	TOTAL
	URBAN	43	7	50
	NON-URBAN	7	50	57
	TOTAL	50	57	107
Total Overall accuracy				0.869

3.4. Discussion

By pre-processing Sentinel 1 images, it was concluded that the VV polarization band yielded higher intensities and amplitudes than VH polarization in (S1A). The VH band for the S1B was observed to have much higher intensities than the VH band in S1A images (The VH polarization sigma^0 in the S1B showed a value of (0.059), much higher than the S1A VH channel (0 .0005)). Both the intensity and amplitude were carried on S1A, S1B generated an equal maximum value of (sigma^0) for the VV bands (0.4). by combining both VH bands of S1A-1B satellites, more information for the built-up regions and urban footprints can be extracted (generated 18 bands).

The speckle characteristics of the stacked image can provide a texture layer that highlights and distinguish built-up surfaces over vegetation and water surfaces. These variations of structures reflectivity led to a local heterogeneity that gives urban areas a very specific and distinct appearance represented in white pixels.

in addition, Using the Gray-Level Co-Occurrence Matrix (GLCM) texture analyses for the sentinel 1A,1B images produced a way better homogeneous values which adds to the speckle divergence step in extracting more information's for the small urban clusters, or settlement areas of differing scales, e. g. small villages as well as cities.

For the overall accuracies of (SAR) images classifications, Sentinel 1A generated a minimum value (OA 0.7666), when classified with (RF) algorithms. however, it failed in detecting urban areas with (ML) classification algorithm (OA 0.599). The combination method of stacking S1A-S1B satellite images produced high overall accuracy (OA 0.803) when classified with (ML) algorithms (generated by the error pixels count on Scp plugin Qgis). and when compared to S2 assigned point of the stratified random sampling model Reaching (OA 0.869). in (RF) classification algorithms for the combination, it produced low accuracy of (OA 0.566), the results of the classification are shown in Table 8. which proves that the hybrid methodology used in the study have a lot of future potentials in discovering built-up environment with Maximum likelihood (ML) classification algorithms.

Table 8: Overall accuracies of the classification's scenarios.

***Satellites* (OA)**	**Supervised *Classification algorithms***	
	Maximum likelihood (ML)	Random forest (RF)
Sentinel 1A	0.599	0.7666
Sentinel 1A/B	0.803	0.566

4. Conclusion

The use of (SAR) satellites provides the possibility to explore urban lands, without relying on weather conditions or the presence of adequate lightning, in addition, the results classifying these images showed that high levels of accuracy in the selection of urban areas can be achieved using the process of combining satellite images of Sentinels satellite. However, Sampling in SAR data needs to be over larger areas to ameliorate speckle and better establish a local 'normal' value of backscatter values.

The combination techniques illustrated in this paper, utilizing the (SAR) Sentinel 1A and 1B discovered built-up areas with Overall accuracies (OA) range from 80% to 86%, which proves to be a reliable method in future studies for built-up environment exploration.

5. Recommendations

The results of the study indicate that making full use of spatial information is important to improve the process of exploring land cover classification. Given that urban footprints maps are characterized by large uncertainties, using RS satellite images offers a much more cost-effective approach to spot informal residential complexes need to be relocated. in addition, Incorporation of suitable textural and spectral images and the use of segmentation-based classification methods can significantly improve land cover classification comparing with traditional per-pixel spectral based classification methods. We also recommend that the methodology followed, should guide the study, and to be linked to the strategic plan for informal housing expansions limitation strategies in the (ARE).

6. Acknowledgement

The authors wish to acknowledge the United States Geological Survey (USGS) for providing the data used for this study, also all the open-source platforms used in the study (QGIS, ESA Sentinel toolbox platforms).

7. References

[1] Bailang Yu, e. -a. (2014). Object-based spatial cluster analysis of urban landscape pattern using nighttime light satellite images: a case study of China,. Geographical Information Science, 2328-2355. doi: DOI: 10.1080/13658816.2014.922186
[2] C. Small A. O. 2018 Remote Sensing of Urban Environments. Comprehensive Remote Sensing Elsevier 96-127 doi: https://doi.org/10.1016/B978-0-12-4095489.10380.
[3] Climate data org 2021 August 28 Retrieved January 1 2021 From https://en.climate-data.org/africa/egypt/damietta-governorate/dumiat-3628/:https://en.climate
[4] Cochran W. G. 1977 Sampling techniques. John Wiley & Sons.
[5] ESA Retrieved August 29,2021, from ESA: sentinels.copernicus.eu/web/sentinel/user-guides/sentinel-1-sar/product-types-processing-levels
[6] Filipponi F 2019 Sentinel-1 GRD Pre-processing Workflow. MDPI, Istituto Superiore per la Protezione e la Ricerca Ambiental (3rd International Electronic Conference on Remote Sensing), proceeding. Retrieved from https://www.researchgate.net/publication/333695202_Sentinel-1_GRD_Preprocessing_Workflow
[7] G Congalton R 1991 A review of assessing the accuracy of classifications of remotely sensed data. Remote Sensing of Environment, 37 35-46 doi: https://doi.org/10.1016/0034-4257(91) 90048-B
[8] Gómez M G 2017 Joint use of Sentinel-1 and Sentinel-2 for land cover classification: A machine learning approach. Sweden: Department of Physical Geography and Ecosystem Science. Retrieved from http://lup.lub.lu.se/luur/download?func=downloadFile&recordOId=8915043&fileOId=8915048
[9] Hashemian et al 2001 Study of Sampling Methods for Accuracy Assessment of Classified Remotely Sensed Data. (A. A. Mohammad S. Hashemian, Ed.) Proceedings of the 20th International Society for Photogrammetry. Retrieved from https://www.semanticscholar.org/paper/STUDY-OF-SAMPLING-METHODS-FOR Accuracy-Assessment Hashemian-Abkar/0fac07aef155bfae046e21ebb7d7f50b612ec168
[10] Kumar D 2021 Urban objects detection from C-band synthetic aperture radar (SAR) satellite images through simulating filter properties. Gautam Buddha Nagar: Amity Institute of

Geoinformatics & Remote Sensing (AIGIRS), Amity University. doi:https://doi.org/10.1038/s41598-021-85121-9

[11] Makinde et al 2019 Land cover mapping using Sentinel-1 SAR and Landsat 8 imageries of Lagos State for 2017. (&. E. Esther Oluwafunmilayo Makinde, Ed.) Environmental Science and Pollution Research, 66–74. doi:https://doi.org/10.1007/s11356-019-05589-x

[12] Martino Pesaresi et al. (2016). Assessment of the Added-Value of Sentinel-2 for Detecting Built-up Areas. Retrieved from https://www.mdpi.com/2072-4292/8/4/299/htm

[13] MathWorks I Retrieved October 1, 2020, Texture Analysis Using the Gray-Level Co-Occurrence Matrix (GLCM) From Texture Analysis Using the Gray-Level Co-Occurrence Matrix (GLCM) Retrieved from https://www.mathworks.com/help/images/texture-analysis-using-the-gray-level-co-occ:https://www.mathworks.com/help/images/texture-analysis-using-the-gray-level-co-occ

[14] Methanex Corporation middle east/Egypt/about-Methanex-Egypt. (emethanex) Retrieved September 1, 2018, from https://www.methanex.com/location/middle-east/egypt/about-methanex-egypt

[15] QGIS 2020 - QGIS Official Website Retrieved November 27, 2018, https://www.qgis.org/en/site/

[16] Semenzato Et Al 2020 Mapping and Monitoring Urban Environment through Sentinel-1 SAR Data: A Case Study in the Veneto Region Italy ISPRC 9-375 doi: ISPRS Int. J. Geo-Inf. 2020, 9, 375; doi:10.3390/ijgi9060375

[17] Shawkat Y Marsadomran Retrieved 2020 10 12 From http://marsadomran.info/en/policy_analysis/2020/01/1880/#notes_and_references

[18] Tahoun et al 2016 Satellite Image Matching and Registration: A Comparative Study Using Invariant Local Features (M. T. Hassanien, Ed.) Image Feature Detectors and Descriptors, 135-171. Retrieved from https://link.springer.com/chapter/10.1007%2F978-3-319-28854-3_6#citeas

[19] Tomás R 2014 Spatiotemporal characteristics of the Huangtupo landslide in the Three Gorges region (China) constrained by radar interferometry Geophysical Journal International, Volume 197, Issue 1, April 213–232 Retrieved from https://academic.oup.com/gji/article/197/1/213/692443

[20] UN 2020 Retrieved November 27, 2021. European Global Navigation Satellite System and Copernicus: Supporting the Sustainable Development Goals - Building Blocks Towards the 2030 Agenda. Vienna: Herausgegeben von United Nations Publications. doi: ISBN: 9211303958

[21] Wegmüller U et al 2015 On the Estimation and Interpretation of Sentinel-1 TOPS InSAR Coherence Proc. of FRINGE. (S. M. Wegmüller U, Ed.) ESA SP-731

"Constructed Wetland Park as Happy Public Space to Achieve Quality of life: Case Study of 10 Ramadan City"

M EL Barmelgy [1], S M Elsaman [2], D T Refaat [2], A Haron[3] and A E El Maalt [4]

[1] Professor, Department of Architecture, Faculty of Engineering Cairo University, Egypt
[2] Postgraduate student, Department of Architecture, Faculty of Engineering Cairo University, Egypt
Email:arc.shmagdy@gmail.com, Darintarek96@gmail.com
[3] Assoc. Prof, Department of Architecture, Faculty of Engineering, CHI Academy, Egypt
Email:Archharon@gmail.com
[4] Teacher, Department of Architecture, Faculty of Engineering Cairo University, Egypt

Abstract. Happy public park (HPS) spaces can be defined as a place that gives us a positive perception that motivates us to spend time and live an unforgettable experience in our memory. As important as the urban blue-green spaces, constructed wetland parks (CWPs) have a spread of environmental and social benefits. Constructed wetlands now are used as an environmental contrivance in many countries. It achieves many benefits like increasing biodiversity, habitats, water treatment, and decreasing pollution. In the last decade, this technology was used in Egypt as a tool for water treatment-funded projects like Manzla lakes and small-scale projects.
The research aim is to identify and define the relation between a happy public space (HPS) and a constructed wetland park (CWP) and evaluate the indicators which can create a constructed wetland park (CWP) as a happy public space (HPS), The simulated data collected from an experts' point of view. The method is to link the indicators that affect the public space especially in a Case study on the 10th of Ramadan "constructed wetland park" to be happy public space and analyzing (WTP) and (HPS) indicators through interviews, workshops, and questionnaires as were statistical analysis of the linkages between the indicators to measure the effective indicators and removing ineffective indicators, to reach a guideline which can create a happy public space in constructed wetland park.

Keywords: *happy public space– constructed wetland park- New city – 10th of Ramadan- quality of life.*

1. Introduction

Public Parks are in which human beings and different social corporations participate (Hajar et.al,2001), and that they make contributions to the wholesome improvement of the city, which might be the critical components of city areas to enhance the excellent of lifestyles of a society. Constructed wetlands structures are mainly appropriate for small groups in improvement, it has proved to be an appealing and strong opportunity due to their low value, and power savings [1]. Also, there may be the benefit of multi-reason re-use of the exquisite discharge, self-remediation, and self-version to the encompassing situations and surroundings [2]. So if we describe the CWPs as HPS we must ask a few questions what are the HPS? and how a CWPs is an HPS? This query must be asked before the collection of any information and making analysis. many solutions will be fined whilst going through this question, and some of the imperative solutions you're much more likely to discover include: "The places used by the public and those are regularly termed in public-owned locations, those are exciting and without problems available to the network and are freed from value and no earnings motive is there at the back of the making plans of those locations [3]. to construct an surroundings to turn out to be tailored as social values evolve in response to converting economic institutions and lifestyles [4]. the purpose of this paper is to evaluate the criteria and indicators in designing CWPs as a HPS. By analyzing the indicators of a HPS and CWPs in a case look at at the tenth of Ramadan, the simulated information have been gathered from experts' interviews, workshops, and questionnaires.

2. Literature Review

2.1. Public Space define (Ps) , values and types

Space that is open and accessible to the people is known as a public space [5]. There are so many things included in the public spaces, such as pavements and roadsides, gardens and parks, public squares, and beaches included in public spaces [6].

The public space is a word which is sometimes misconstrued to mean a different kind of places for architecture such as a space to meet up which is among the social space's broader concept [7]. For these type of public spaces commons is one of the earliest examples

2.1.1 Public spaces (Ps) – (re)defined

Thus, it is difficult to define the public domain as a place. Many types of public space –quite varied on the face of it. So, it is misleading that something is called the public space as it stands. With all forms of qualifications, the public spaces are those where we move around and interact with other people [8]. The public places are an abstract ideology covering the fields where everyone agrees to interact.

- As urban civility banners the use of public spaces
- These public spaces also serve as the urban commons
- Income, investment, and creation of wealth or revenue is promoted with the help of these spaces.
- The environmental sustainability is enhanced as well by these public spaces
- These urban public spaces are likely to have a significant impact on transport efficiency.
- Public health is also improved with the help of these public spaces
- These public places ensure urban safety.
- Public spaces promote equity and social inclusions.
- The public places promote an age-friendly and gender-friendly environment
- The use of these public spaces promotes citizen involvement in city affairs
- Public spaces are the keys to make great cities.

2.2. Happy public space (HPS)

The definition of urban happiness is emphatically interlaced with social, natural, economic and philosophical thinks approximately and, in settlement with the short trade of methods of life, wishes and propensities, the definition is inside the put together of nonstop alter [9]. So, it's a set of human beings and framework, be that because it may, there is a well-advanced framework in positioned to assure the well-being, comfort and exchange of thoughts, unmistakable and intangible views of a positioned and the sporting activities finished via way of means of the folks who stay and make use of them. What may be visible via way of means of the eye, heard via way of means of the ear, or a scene it really is an come upon in itself and a reminiscence and as a result making a valid enthusiastic framework, known as a happy city [10]. On the opposite hand, the definition suggests as much as be comparative to that applied for characterizing well-being, exceptional of life, and, in a sure sense, supportability [11]. More organizations of making happy urban spac`e according table(1)

organization	Definition organization	Domains and indicator
Happy city	The Happy City Approach is associated with the technological know-how of environmental psychology and has resulted from social psychology studies since the 1970's with Milgram's look at of the enjoy of living in cities [12]	Eight principles are key element sociability, comfort, joy, healthy, Equidae, resilience, meaning, belonging.
Project of Pub Space(pps)	Is a nonprofit corporation devoted to assisting human beings create and preserve public areas that construct sturdy communities, Founder through Fred Kent, 1975, Headquarters in New York, United States [8].	four qualities: accessible; people are engaged in activities there; the space is comfortable and has a good image; and finally, it is a sociable place
GNH Index	It is a dimension device used for coverage making to growth GNH. The GNH Index is likewise called the GNH Happiness Survey. It consists of the 9 domain names that are in addition supported via way of means of the 33 indicators, 2008,Thailand [13].	9 domains Living standards, Education Health, Environment, Community, Vitality Time-use, psychological well-being Good Governance, Cultural resilience and promotion.
Cities in Moti Index dimensio	The IESE Driving Cities Index for Social Enterprise Improvement and Efficiency is a take a look at the usage of publicly to be had data, a surveyed index 148 cities in fifty seven nations for its 2015 report, fifty five of which might be capital. [14]	9 indicators, human, capital, socal chosen, economy, public management, governance environment, mobility, urban planning, international outreach, technology.
Green City Ind	Created in 2009, the Green City Index is a tool to assist towns examine their environmental performance , The advantage of this method is to lessen environmental impacts. [15]	8 index, west land use, water, building, co2, energy, environment governance, air quality, transport.
Global Comp Cities Circles Sustainability	Circles of Sustainability technique facilitates city grou apprehend the way to achieve sustainability. It has be used by many cities, amongst them Johannesbu Melbourne, New Delhi, Sao Paulo and Tehran [16].	four domains: ecology, economics, politics and culture. Each of these domains is divided in seven subdomains

table 1: organizations of making happy urban space, By authors

2.3. Construction wetland

constructed wetlands are man-made wetlands, simulating the functions of natural wetlands, the primary cause of which is water purification. The constructed wetland, a typically constructed ecosystem, was initially developed about forty years ago in North America and Europe to utilize and recover the biodegradation capacity of plants. Possessing the blessings of low production and operating costs as well as its capacity to be used alone or with other systems. Constructed wetlands' systems are especially suitable for small communities in developing nations where potential health benefits from pathogen removal are considerable [17]. Constructed wetlands for wastewater treatment can be labeled in line with the life shape of the dominating macrophyte, into systems with free-floating, floating leaved, rooted emergent and submerged macrophytes. They can be labeled additionally in line with the hydrology of the wetland and subsurface flow [18].Constructed wetlands have proved to be an appealing and solid opportunity due to their low cost, and power savings. Also, there may be the gain of multi-cause re-use of the high-quality effluent, self-remediation, and self-adaptation to the encircling situations and environment.[2].

2.3.1 Constructed wetland parks:

Constructed wetlands are increasingly more called as a substitute low-cost, energy-efficient, natural techniques of treating sewage, industrial and agricultural wastes, and storm water runoff at the same time as concurrently supplying the opportunity for numerous benefits[19]. Constructed Wetland Park gives the capability for integration of constructed wetlands into park and leisure systems, the wildlife habitat they offer, their aesthetic values, and the greater quality effluent that they deliver, which can be recycled for panorama irrigation or impounded in an appealing and academic pond of value in attracting wildlife at the same time as also conveying information on wetlands procedures[18].

IOP Conf. Series: Earth and Environmental Science **992** (2022) 012009 doi:10.1088/1755-1315/992/1/012009

2.4. New Cities in Egypt

The establishment of latest towns in Egypt changed into accomplished in successive stages main to the emergence of 3 generations of the brand new towns as follows: First generation: sixth October City, Tenth of Ramadan, Sadat City,15 May City, New Borg El Arab, New Salehia, New Damietta and Touristic Villages, Second generation: Sheikh Zayed City, Oubour City, Badr City, New Noubarya, New Beni Suef and New Minya ,and Third generation: Rehab, El Sherouk , New Cairo, New Assuit, Tiba, New Sohag, New Aswan, New Qena, New Fayoum and New Akhmim.five The simple issues in the back of the introduction of latest towns range in line with the numerous structures and social, monetary and environmental conditions. At the identical time, those towns are taken into consideration as a mirrored image of positive concepts [20].

2.4.1. 10th of Radman city

proposed in the late 1970s. Appendix presents a summary of the current situation of the cities in 2010, the time when the strategic master plans studies took place. it share the fact that the industrial activities are the main drivers of the economy and they are located more or less at equal distances from Cairo, one at the East and the other on the West. Its achieved successful industrial development while their population failed to meet the targeted number in the original master plan. Through a review of the basic information of both cities, their current situation and then the proposed vision, objectives and development strategies proposed, a number of observations can be drawn.

-10th of Ramadan is one of the first New Cities planned in Egypt and have developed over the past 30 years as the largest industrial cities in Egypt (e.g., 10th of Ramadan has about 1400 factories, annual production of more than Billion EGP 75.42, employing approximately 188,166 workers) [20].

- Both city suffer from serious issues related to housing, environment, urban/civic life.

The root causes point mainly at poor planning, implementation and weaknesses in legal/institutional frameworks.

3. Case study 10th of Ramadan constructed wetland park

The main idea of the project is the design and construction of a multifunction landscape project which sustainably combines landscape features and ecological functions.

In this phase, studies have been conducted on several aspects such as design, environmental and economic aspects. The aim of this phase was to instigative potentials and seeks best opportunities. The relation between our site and its surrounding was our major scope of study.

Figure 1: 10th of Ramadan constructed wetland park By authors

3.1 Design Zoning:

Zoning phase concluded all our previous studies and shed light on land potential and activities that can be made to serve the park. Land was divided into 5 sectors 200m each, the wetland area which is between sector 5 and 3 with consideration of water intake from water treatment plant.

Sector 1 and 2 shows the area of the main park which will include most of activities. design concept was how to convert the site proportion from this static shape to a dynamic form.

4 . Research Methodology:

This research aims to achieve the guideline of design (CWP) as a happy public space in order to provide a positive contribution to urban design public parks.

This research consists of four-phase:

- The first phase is Analyzing Theoretical data and Case studies for a happy public space (HPS) to combine Indicators of (CWP) and happy public space (HPS).

- In the second phase Expert interviews entails observing stakeholders' and experts' opinions through two stages 1st Stakeholder's interview,2nd Stockholder's workshop, and 3rd Questionnaire which concern Built environment, social environment, and economy which contribute to the feeling of happiness.

- The third stage, used (Multiple questionnaires) as a tool for measuring people opinions using google forms (questionnaire form,2021), the questionnaire is divided into four main core parts, interviewee background, indicators of built environment factors, indicators of social environment factors and indicators of economic factors.

- The fourth phase, Integration Between the first phase and the second phase to achieve an effective Guideline for a happy public wetland park (HPWP).

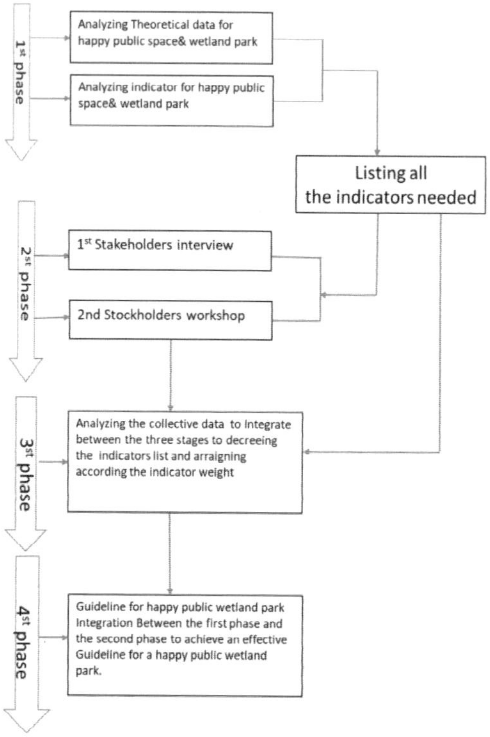

Figure 2: Research Methodology
Source: by authors

5. finding and Results

5.1 The first phase

The first phase is Analyzing Theoretical data and Case studies for a happy public space (HPS) and Constricted wetland park (CWP) to combine Indicators of (CWP) and (HPS).

5.1.1. A happy space is set by many factors which are in turn influenced by a variety of indicator

Built Environmental	Social Environment	Economic
• Facilitate a healthy life • livability, felling • comfortable • Helping with long-term growth • cultural facilities • recreation facilities • aesthetics of place, • Facilitate a healthy life • quality of space ,	• Develop strong social relationships • Affiliation of the place • Involvement • Achieving the human instinct • Openness • Diversity • Spread the happiness • connectedness & networking.	• Achieving the human instinct • creative outcome • investments • Realizing opportunities and public resources • Encouraging economic diversification.

Table 2: Happy space indicator, By authors

5.1.2 Constricted wetland park is set by many factors which are in turn influenced by a variety of indicator

According to the theoretical data and case study analysis has shown that application of constructed wetland parks is a good and economic, Social, and environmental tool to achieve sustainable development goals. These indicators concluded from all the studied projects, as shown in table (3) The more than 20 national and international case studies that has been Analyzed with a specific criteria like (Sydney Park, . Shanghai Houtan, Renaissance Park, Bahr El-Baqr Constructed Wetland, Wadi hanifa ….etc.).

IOP Conf. Series: Earth and Environmental Science **992** (2022) 012009 doi:10.1088/1755-1315/992/1/012009

Built Environmental	Social Environment	Economic
• Improving air quality • Spatial Diversity • Cultural facilities • Recreational facilities • Healthy lifestyle • Energy production • Reducing urban heat island • Reducing noise pollution	• Social interaction • Enjoy nature • Healthy life style • Improve aesthetics • Sense of belong • Connectedness • Diversity of cultural and • creative class • Spatial Diversity • Flexibility of public space.	• Provide job opportunities • Increase property • Increase property Value • Use pf recycling • Save in water treatment • Touristic value • Leader ship and entrepreneur ship • Prices of accommodation facilities • Cultural employment and the creative

Table 3: Constricted wetland park indicator, By authors

5.2. second phase:

In the second phase Expert interviews entails observing stakeholders' and experts' opinions through two stages 1st Stakeholder's interview,2nd Stockholder's workshop.

5.2.1 stage 1: Expert interviews, factor list of main issues and problems

stage is an interview with Stockholders and the researcher gathering meeting and describing in a short demonstration about the construction wetland park(CWP) and happy public space(HPS), attending 20p, the outputs of this interview are:

Issues & problem	Description
Making space	placemaking inspires people to collectively reimagine and reinvent public spaces as the heart of park.
Quality of space	quality spaces understand how they achieved and designed and meeting the challenges of social, economic and environmental demands.
The cost of establishing gardens	How to design with low cost
garden management	responsible for knowing everything about garden plants, mechanical systems, and landscape structures to be able to provide you the best landscape maintenance services.
Outcome	How to increase the garden's income to spend on itself

Table 4: Constricted wetland park indicator, By authors

5.2.2. stage 2 Expert workshop:

A stakeholder's workshop was held in Cairo on 6, 7 February 2021. that workshop invited academics, the city management team, designers, residents, researchers, post-graduate students, attendance was more than 80 persons and the output s of this workshop is some proposals table(5)

Proposals	**Description**
interesting	Different places for social interaction between age groups to raise the level of belonging to the place
Network interaction	Provides an online social network
activity	Creative activities for all ages
More green space	The green color gives a sense of comfort in the place
Leadership and entrepreneur ship	Establishing a different advertisement for the project to attract Leadership and entrepreneur ship
Recycling	Using recycling in furniture to make difference and save cost

table 5: Expert workshop output, By authors

IOP Conf. Series: Earth and Environmental Science **992** (2022) 012009 doi:10.1088/1755-1315/992/1/012009

5.3. Merge first phase & second phase

to combine Indicators of (CWP) and (HPS). the outputs are:

Dimensions	**Indicators**	**Description**	**Reference (based on)**
Environmental	Water quality	a healthy life is improved with the help of water quality in public space	[18],[15],[6]
	Comfortable	Feeling comfortable while using the space (visual -spatial use-......)	[8],[19],[12]
	Energy reducing	Preserving the natural resources in the surrounding environment	[18],[3],[19]
	Culture facilities	Cultural diversity adds weight and importance to the place	[14],[12],[8]
	Spatial diversity	Spatial diversity gives a sense of lack of boredom and enjoyment	[18],[8],[12],[14]
	More green space	The green color gives a sense of comfort in the place	Work shop
	Livability	Accept the place and the desire to sit in it for more time	[12],[8]
Social	Social interaction	Creating spaces for social interaction	[8],[12],[14]
	Diversity activity	The variety of activities helps to didn't feel bored	Work shop, interview
	Enjoy Nature	Feeling comfortable and merging with nature	[15],[8]
	Sense of belonging	A sense of belonging helps preserve the place	[18],[12],[13]
	Active attractive	The presence of attractive activities creates happiness for those who use the space	[13],[8],[12]
	Ecology participation	The interference of natural elements in the space gives a sense of comfort	[18],[16]
	Culture and art interactions	To create a spirit of creativity and diversity among the users of the space	[8],[12],[14]
	Network interaction	Provides an online social network	Work shop
Economic	interesting	Different places for social interaction between age groups to raise the level of belonging to the place	Work shop
	Creative out come	Different ideas to increase the income of the park to spend on itself	Work shop, interview

	Tourist attraction	Encouraging different types of park to diversify and attract internal and external tourism	[18],[15],[6]
	Leadership and entrainer ship	Attract Leadership and entrainer ship gives more opportunities to increase income	Work shop
	Price od accommodation facilities	The park 's success helps raise its accommodation facilities value	[18],[15],[6]
	Water Treatment saving	Wastewater desalination and its use in agriculture	[18],[15],[6]
	Provide job opportunity	Opening new projects creates job opportunities	[18],[15],[6]

Table 6: Indicators of (CWP) and (HPS), By authors based on theoretical, interviwe, workshop

5.4. Third phase

In the third phase, using Multiple questionnaires as a tool for measuring people opinions using google forms (questionnaire form,2021), the questionnaire is divided into four main core parts, interviewee background, indicators of built environment factors, indicators of social environment factors and indicators of economic factors. The number people who participated in the survey was 100 and divided as follows:

sample	specialist	architects and professor	Ph.D. researcher	MS.C researcher	post -graduate students
%	20%	23%	18%	22%	17%

Table 7: The number of people who participated in the survey, By authors

5.5. fourth phase

The results of four phase, aimed to evaluate the stakeholder satisfaction to Evaluate the criteria and indicators in design constructed wetland park (CWPs) as a happy public space (HPS).

55.1 statistical questionnaire

Use the statistical questionnaire conducted as a need of participatory design approach in the context of stakeholder – designer relationship at the design phase

5.5.1.1 The stability coefficients for the indicators

It is clear from the previous table (8) the values of Cronbach's alpha stability coefficients for the dimensions of the study, all of which were greater than (.70), where the value of the stability coefficient for the Built Environment axis was 0.939. This indicates that if the question of the sample members is repeated again, we get the same response by 93.9%, which is a degree of Very high stability, and the validity coefficient reached 0.969, which is a high degree of reliability indicating a good understanding of the items of the Built Environment axis by the sample members.

- The reliability coefficient of Cronbach's alpha for the Social Environment axis was 0.973, and this indicates that if the question of the sample members was repeated again, we would get the same response with a rate of 97.3%, which is a very high degree of stability, and the validity coefficient reached 0.986, which is a high degree of credibility indicating a good understanding of paragraphs Social Environment axis by respondents with a percentage of 98.6%.
- The reliability coefficient Alpha Cronbach for the Economic axis reached 0.953, and this indicates that if the question of the sample members is repeated again, we get the same response by 95.3%, which is a very high degree of stability, and the validity coefficient reached 0.976, which is a high

degree of credibility indicating a good understanding of the paragraphs of the axis Economic by 97.6% of respondents.

From the foregoing, the results indicate that it is possible to rely on those obtained opinions, and then adopt the results of characterizing the sample trends.

	N of Items	Cronbach's Alpha	Validity
Built Environment	7	0.939	0.969
Social Environment	9	0.973	0.986
Economic	8	0.953	0.976

Table 8: the values of Cronbach's alpha stability in the survey, By authors

5.5.1.2. Evaluate the criteria and indicators

The validity of the factor analysis means the extent to which each paragraph of the survey is consistent with the field to which this paragraph belongs. The extent to which the paragraphs are saturated is measured on the total axis of the scale to which they belong. Factor analysis was used with a focus on some measures such as the adequacy of the sample KMO and Bartlett test and the interpreted variance ratio AVE.

The validity of the internal consistency was tested using the factor analysis test, to find out the most important statements that the axis could contain and with which the test assumptions were verified, in addition to the use of neural networks to determine the relative importance of each paragraph.

- *Built Environment*

By studying the factorial analysis of those items and their representation of the Built Environment axis, the results showed that the sample adequacy test coefficient KMO reached 0.910, which is greater than 0.50. The results also showed the significance of the Bartlett test (Chi Square) with a degree of freedom of 21 at a confidence level of 99%, where the significance of the test 0.000, which is less than the significance level of 1%, which indicates that there is a correlation between the phrases and each other, and therefore the factor analysis test must be conducted. The loading coefficient for each of the axis's paragraphs ranged between 0.613 and 0.935 and all of them weregreater than 0.50. Therefore, these paragraphs strongly express the Built Environment axis and that none of these paragraphs can be deleted.

		Component
Comfortable place	X1.8	0.893
More green space	X1.9	0.815
Culture facilities	X1.10	0.907
Energy reducing	X1.11	0.613
Livability	X1.12	0.882
quality to a healthy life	X1.13	0.935
Spatial diversity	X1.15	0.923
Kaiser-Meyer-Olkin Measure Of Sampling Adequacy.		0.910
Approx. Chi-Square		606.657
Df		21
P_Value		0.000
AVE		73.735

Table 9: the factorial analysis representation of the Built Environment axis survey, By authors

table No. (10) shows the relative weight of the importance of the paragraphs within the Built Environment axis. It was found that paragraph x1.13 is the most important among the paragraphs of the axis with a relative importance of 19%.

		Factor Score	Nueral Network Score	Rank
Comfortable place	x1.8	0.173	0.130	6
More green space	x1.9	0.158	0.131	5
Culture facilities	x1.10	0.176	0.137	3
Energy reducing	x1.11	0.119	0.135	4
Livability	x1.12	0.171	0.147	2
quality to a healthy life	x1.13	0.181	0.190	1
Spatial diversity	x1.15	0.179	0.129	7

Table 10: the relative weight of indicator, By authors

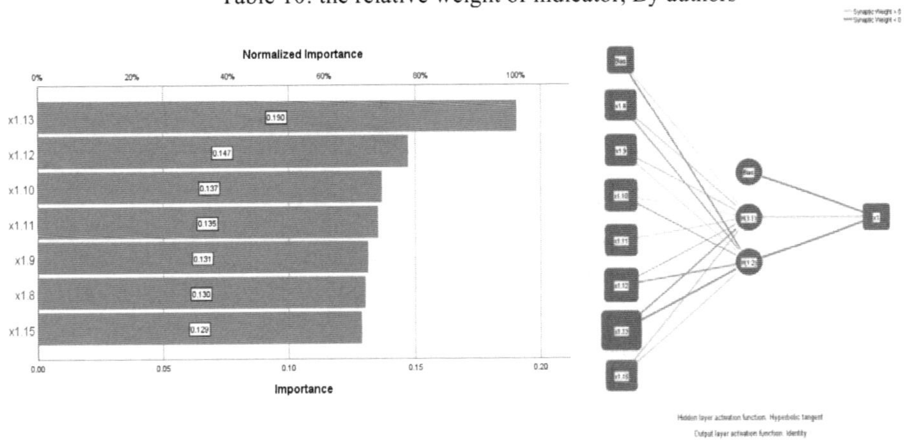

Figure3: the relative weight the Built Environment axis
Source: by authors

		Component
interesting	x2.16	0.926
Network interaction	x2.17	0.868
Culture and art interactions	x2.18	0.921
Sense of belonging by activities	x2.19	0.912
Active attractive	x2.20	0.936
Ecology participation	x2.21	0.856
Enjoy Nature	x2.22	0.888
Diversity activity	x2.23	0.893
Social interaction	x2.24	0.960
Kaiser-Meyer-Olkin Measure of Sampling Adequacy.		0.917
Approx. Chi-Square		1105.468
df		36
P_value		0.000
AVE		82.305

Table 11: the factorial analysis representation of the social Environment axis survey, By authors

- ***Social Environmental***

By studying the factorial analysis of those items and their representation of the Built Environment axis, the results showed that the sample adequacy test coefficient KMO reached 0.917, which is greater than 0.50. The results also showed the significance of the Bartlett test (Chi Square) with a degree of freedom of 36 at a confidence level of 99%, where the significance of the test 0.000, which is less than the significance level of 1%, which indicates that there is a correlation between the phrases and each other,

and therefore the factor analysis test must be conducted. The loading coefficient for each of the axis's paragraphs ranged between 0.856 and 0.960 and all of them weregreater than 0.50. Therefore, these paragraphs strongly express the Built Environment axis and that none of these
table No. (12) shows the relative weight of the importance of the paragraphs within the Built Environment axis. It was found that paragraph x2.24 is the most important among the paragraphs of the axis with a relative importance of 15%.

		Factor Score	Nueral Network Score	Rank
interesting	x2.16	0.125	0.103	4
Network interaction	x2.17	0.117	0.101	6
Culture and art interactions	x2.18	0.124	0.133	2
Sense of belonging by activities	x2.19	0.123	0.099	8
Active attractive	x2.20	0.126	0.103	5
Ecology participation	x2.21	0.116	0.089	9
Enjoy Nature	x2.22	0.120	0.100	7
Diversity activity	x2.23	0.121	0.122	3
Social interaction	x2.24	0.130	0.151	1

Table 12: the relative weight of indecator, By authors

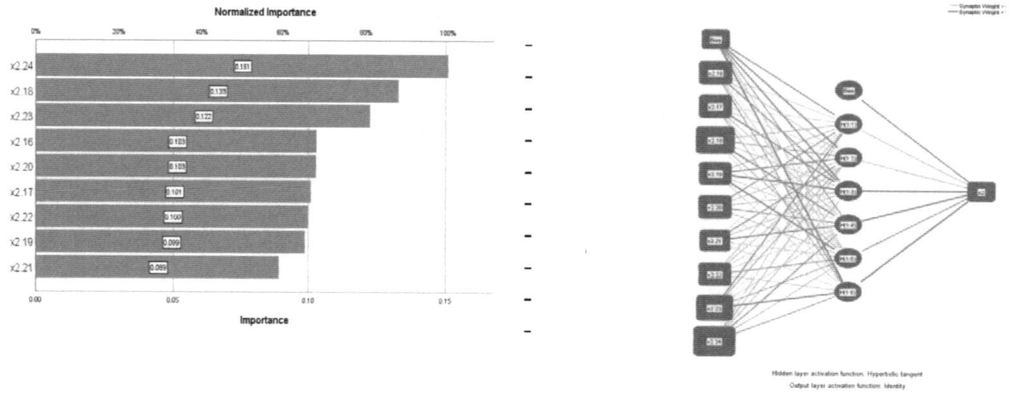

Figure4: the relative weight the social Environment axis
Source: by authors

- *Econmic*

By studying the factorial analysis of those items and their representation of the Built Environment axis, the results showed that the sample adequacy test coefficient KMO reached 0.913, which is greater than 0.50. The results also showed the significance of the Bartlett test (Chi Square) with a degree of freedom of 28 at a confidence level of 99%, where the significance of the test 0.000, which is less than the significance level of 1%, which indicates that there is a correlation between the phrases and each other, and therefore the factor analysis test must be conducted. The loading coefficient for each of the axis's paragraphs ranged between 0.763 and 0.926and all of them weregreater than 0.50. Therefore, these paragraphs strongly express the Built Environment axis and that none of these paragraphs can be deleted.relative importance of 15% paragraphs can be deleted.

		Component
Leadership and entrainer	x3.26	0.848
Creative out come	x3.27	0.921
Tourist attraction	x3.28	0.921
Recycling Furniture	x3.29	0.860
Water Treatment saving	x3.30	0.914
Price od accommodation facilities	x3.31	0.779
Realizing opportunities and public resources	x3.32	0.926
Provide job opportunity	x3.33	0.763
Kaiser-Meyer-Olkin Measure of Sampling Adequacy.		0.913
Approx. Chi-Square		764.654
df		28
P_value		0.000
AVE		75.483

Table 13: the factorial analysis representation of the Econmic

table No. (14) shows the relative weight of the importance of the paragraphs within the Built Environment axis. It was found that paragraph x3.27 is the most important among the paragraphs of the axis with a relative importance of 15%.

		Factor Score	Nueral Network Score	Rank
Leadership and entrainer	x3.26	0.140	0.135	3
Creative out come	x3.27	0.153	0.155	1
Tourist attraction	x3.28	0.153	0.150	2
Recycling Furniture	x3.29	0.142	0.128	5
Water Treatment saving	x3.30	0.151	0.099	7
Price od accommodation facilities	x3.31	0.129	0.107	6
Realizing opportunities and public resources	x3.32	0.153	0.092	8
Provide job opportunity	x3.33	0.126	0.135	4

Table 14: the relative weight of indecator, By authors axis

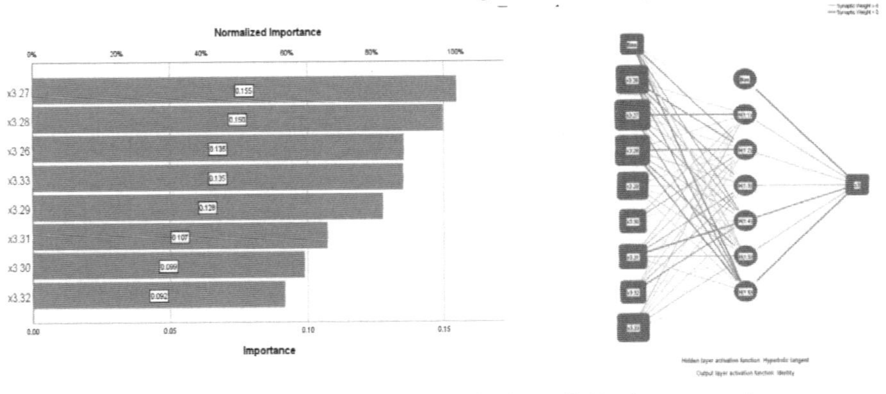

Figure5: the relative weight the Built Environment axis
Source: by authors

6. Conclusion

This research illustrates the value of establishing public parks aimed at increasing the happiness of users of new Egyptian cities. The research proved the important of increasing the public space in new city, especially 10th of ramadan and achieve the needs of users.
The case study proved from point and view stakeholders who participated in the questionnaire that the constricted wetland park successful way to raise the quality of life in Egypt.
From the theoretical and analytical study shows the needs for further in urbanism and landscape researches to conform with Egyptian environment.
Egyptian cities face significant urban and environment challenges, so it requires to linking scientific research to application and integration visions of stakeholders and specialists to reduce the gap between research and application on the reality.
The research output the guideline table (15) framework for establishing theses projects in Egypt to be integrated with the site co-ordination to achieve development goals.

Built Environment	weight	Social environment	weight	economic	weight
quality to a healthy life	%19	Social interaction	%15.1	Creative out come	%15.5
Livability	%14.7	Culture and art interactions	%13.3	Tourist attraction	%15
Culture facilities	%13.7	Diversity activity	%12.2	Leadership and entrainer	%13.5
Energy reducing	%13.5	interesting	%10.3	Provide job opportunity	%13.5
Comfortable place	%13.1	Active attractive	%10.3	Recycling Furniture	%12.8
More green space	%13	Network interaction	%10.1	Price od accommodation facilities	%10.7
Spatial diversity	%12.9	Enjoy Nature	%10	Water Treatment saving	%9.9
		Sense of belonging by activities	%9.9	Realizing opportunities and public resources	%9.2
		Ecology participation	%8.9		

Table 15: the final relative weight of indicators, By authors

7. Acknowledgment

Researcher's thanks, Science Technology & Innovation Funding Authority (STDF), the funder of the project and research to provide all possibilities to carry out this research within framework of the research project.

8. References

[1] Zhang D, Gersberg R. M., Kea T. S., (2009) Constructed wetlands in China, Ecological Engineering 35 pp.1367–137.
[2] urban design compendium, Llewelyn-Davies, Yuriko Sawatani, Tatsunori Hara, Springer Japan 2016.
[3] Services G, Public Life booklet, , Published on Jun 14, 2018.
[4] Hantala M, Urban space as element in urban design with reference to Scottish new town, adissertationprepared for a master in Architecture by, Mackintosh School of Architecture University of Glasgow ,2018.

[5] kurniawati W, Public Space for Marginal People, Procedia-Social and Behavioral Sciences 36 (2012) 476 – 484.
[6] https://www.pps.org
[7] Montgomery J, Making a City: Urbanity, Vitality and Urban Design, Journal of Urban Design, Vol. 3, No. 1, 1998.
[8] Kyttä M, Broberg A, Haybatollahi M, Thomé K, Urban happiness: context-sensitive study of the social sustainability of urban settings , Environment and Planning B: Planning and Design 2016, volume 43, pages 34– 57.
[9] Montgomery C, Happy City: Transforming Our Lives Through Urban Design, ,2013, Canada.
[10] Ballas D, What makes a 'happy city'? Department of Geography, University of Sheffield, Winter Street, United Kingdom ,2013.
[11] Sepe M., Int. J. Sus. Dev. Plann The role of public space to achieve urban happiness,. Vol. 12, No. 4 (2017) 724–733.
[12] https://thehappycity.com/project.
[13] https://www.gnhcentrebhutan.org.
[14] IESE Cities in Motion Index,2015.
[15] Sustainable City Indexing,2017.
[16] https://www.circlesofsustainability.org
[17] Gaber R ,implementation of constructed wetlands landscape design ,a resilient application, journal of urban research, vol.36, April,2020.
[18] Haron A , Faisal Z, Constructed wetland parks a pathway to sustainability for Cairo, journal of urban research, vol.36, April,2020.
[19] Haron A , Faisal Z, multifunction green infrastructure, towards upgrading urban ecosystem in cairo,2017.
[20] http://www.newcities.gov.eg

AUTHOR INDEX